GLOBAL WARNING . . . GLOBAL WARMING

GLOBAL WARNING . . . GLOBAL WARMING

Melvin A. Benarde, Ph.D.

Professor, Center for Environmental Studies,
Temple University

JOHN WILEY & SONS

New York • Chichester • Brisbane • Toronto • Singapore

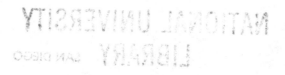
Library of Congress Cataloging in Publication Data:
Benarde, Melvin A.
    Global warning—global warming / Melvin A. Benarde.
      p. cm.
    Includes index.
    ISBN 0–471–51323–7
    1. Global warming. 2. Climatic changes. I. Title.
QC981.8.G56B44 1992
363.73'87—dc20   91-22276
                                          CIP

Printed and bound in the United States of America by Braun-Brumfield, Inc.

10 9 8 7 6 5 4 3 2 1

For Erica Fass Spiritos
My newest grandchild who could come of age in a climatically altered period . . . one in which she might be forced to contend with the type of world described by John King in 1595:

Our years are turned upside down; Our summers are no summers
and our harvests are no harvests.

But if we prove to be an attentive and reasonable people, this will remain only a verse on an ancient page.

PREFACE

I've tried to write a book about an inordinately complex subject in a manner at once highly accessible yet uncompromising in its scientific accuracy; a book about which readers might say "finally I understand the issues," or "I've got it, I know what they're talking about." Were I to hear those types of remarks, the effort would have been well worth it. For an effort it was. Why? Because the subject is vast in its dimensions. It literally takes on the planet, but is so fraught with uncertainty and creative ruminations on the part of a broad clutch of researchers and interested parties that it can become elusive, unless of course one is prepared to accept at face value whatever is offered. I was not. As I see it, while readers want to know the substantive issues, they also want "the bottom line." As a friend asked me, "Do I have to worry?" To respond to either requires the broadest probing of the subject. But when accomplished, the perspective that emerges is less than clear. In fact, at this stage the issue is as the Scots maintain, unproved. Consequently, a definitive point of view is inappropriate and unwarranted.

At this point, we are in thrall to model predictions, or worse, to what the media have dished up for us—a thin gruel at best. In concept and potential fact, "global warming" is far more than model predictions. Thus I wanted readers to come to grips with the subject's broadest dimensions, its many ramifications wherein models and their predictions play a lesser role. Accordingly, it was necessary to discuss the seasons and the atmosphere, and models of course required delineation insofar as they are the engine driving the issue. But to become mired in myriad detail would be pointless. On the other hand, there was a major point to paint with a broad brush in discussing potential dislocations and advantages of a high (or higher) CO_2 world. Here indeed was "the beef." Here I strove for a panoramic view to give readers insights into the formidable array of issues, which, in a warmer world, could well impinge on every facet of our lives: this, in addition to the protean research activities in progress literally throughout the world, seeking, often successfully, to mitigate the untoward consequences of warming, should it occur. A similar approach is taken with potential solutions. Here I tried to show that in fact an impressive host of workable, reasonable solutions are available that do not require shutting down the world, as some would have us do. I am, however, enough of a pragmatist to know that heroic efforts will be required to effect changes of behavior, especially of a population, a people who have developed profligate ways.

Thus the book I believed needed to be written and made widely available is neither a polemic nor a defense of a cherished position. Neither does it pretend to be a scientific treatise. My intention was to produce a book that would serve to clarify and place in perspective a fascinating yet obscure subject without oversimplification or avoidance of issues because of their complexity. Hence this is a broad account and an assessment of an all encompassing issue that could easily become the most consuming and challenging problem with which our nation must contend. But I have also tried to show that "we the people, shall not perish from the earth" because we are, have always been, highly adaptive. We continually transform our communities, making them less susceptible to the vagaries of climate. The evidence is all about us, yet we often fail to see it. Crisis and panic are wholly inappropriate. The possibility of warming is an issue with which the country must deal, as it did so remarkably well with the Gulf War. Time is on our side. But it must not be squandered.

Obviously, this book could not have been done alone. I am indebted to the eagerly helpful, always cheerful librarians at Princeton University's Geology, Biology, and Astronomy libraries, The Princeton Public Library, and the librarians of that marvelously conceived Douglas College Library, and to Betsy Tabas who directs Temple University's Engineering, Computer Science, and Architecture Library. She above all will be relieved that this undertaking is over at last. In the process, I have come to admire and respect librarians as never before. The legerdemain they can perform to make an otherwise inaccessible item appear is the stuff of legends. I remain deeply indebted to them.

The manuscript took shape under the keenly watchful eyes of Mary Dunn, Tammy Urian, and Diane Weinberg of Temple University's Information Processing Center, along with my secretary Diane Dymski, who gave up weekends and evenings to assure a worthy manuscript. Then there were the people who suffered—long and most often silently. From my 5-year-old grandson, Zach, who wanted to "bat a few" during baseball season and "kick a few" during football season, I ask understanding and forgiveness for being unavailable. My wife, Anita, cannot be repaid for all the lost weekends, holidays, and so many other times when she had to go it alone. But she was there for me when I needed advice and suggestions. For her there will be a place in heaven. My son, Scott, a wordsmith and feature writer for The Palm Beach Post, needs to be singled out for providing the provocative title for this book. There was general agreement that he was on the mark.

At this point I must raise my glass to the working scientists, the many extraordinary people such as Syukuro Manabe and Jerry Mahlman of the

Geophysical Fluid Dynamics Laboratory/NOAA, Jim Hansen of Goddard Institute for Space Studies, William Gray of Colorado State University, John Imbrie of Brown University, Hugh W. Ellsaesser of the Lawrence Livermore National Laboratory, Charles D. Keeling, Tim Barnett, and others at Scripps Institution of Oceanography, and the good people at The Lamont–Doherty Geological Observatory, Oak Ridge National Laboratories, and National Center for Atmospheric Research, who helped me gather information and explained and discussed the issues when I asked. Again, it must be said that the best and brightest people are working to understand and define the warming problem, and for me it was a joy to have had an opportunity to be associated with them even in so small a way. Nevertheless, whatever help was extended me, the interpretations, conclusions, errors, and omissions are mine alone.

MELVIN A. BENARDE

Princeton, New Jersey
October 1991

CONTENTS

GLOBAL WARNING . . . GLOBAL WARMING

INTRODUCTION

For extensive areas of the world there is a distinct and disturbing possibility that the stable, generally mild world climate is changing. Forecasts suggest a far warmer world in which rising sea levels and drought in some areas, and excessive moisture in others, would precipitate major ecologic and population dislocations, injury, disease, and death for many. Human activity rather than natural changes is seen as responsible. A major open question is: Can human activity or the result of human activity modify natural forces, or is the current concern about an unnatural, untimely warming trend the consequence of unbridled human ego coupled with ignorance of the workings of the natural world?

The question is at once reasonable and pressing, given the potentially noxious consequences for humankind if we do indeed possess such power. That man is a force to be reckoned with is not a new idea. The seriousness of human activity was not lost on 19th century observers of the human scene. Horace Bushnell (1802–1876), a Congregational minister, averred in his sermon "The Power of an Endless Life" that "not all the works, and storms, and earthquakes and seas, and seasons of the world have done so much to revolutionize the earth as man has done since the day he came forth upon it, and received dominion over it." And according to Antonio Stoppani (1824–1894), an Italian man of letters, "the creation of man was the introduction of a new element into nature, of a force wholly unknown to earlier periods." Stoppani maintained that man is a new telluric force, which in power and universality may be compared to the greater forces of the earth. Echoing Stoppani, George Perkins Marsh (1801–1882), author of *The Earth as Modified by Human Activity* (1874), was certain that, where man was concerned, "there is scarcely any assignable limit to his present and prospective voluntary controlling power over terrestrial nature." To be sure this was the late 19th century and man was beginning to flex his muscles. Under his management the landscape was indeed changing. But did these characterizations of the might of human beings fail to appreciate or comprehend the tectonic and astronomical forces pitted against him, as it were, and can these be transcended?

As with all major issues, potential global warming is not new. It has been discussed and written about for over a hundred years. In September 1861, for example, *The London, Edinburgh and Dublin Philosophical Magazine* and *Journal of Science* published a seminal paper by John Tyndall, British physicist and Fellow of the Royal Society, on the effects on climate of the transmission of solar and terrestrial heat through the earth's atmosphere. "Now," he wrote, "if as the above experiments indicate, the chief influence be exercised by the aqueous vapour every variation of this constituent must produce a change of climate. Similar remarks would apply to the carbonic acid diffused through the air . . . . It is not. . . necessary to assume alterations in the density and height of the atmosphere to account for different amounts of heat being preserved in the earth at different times; a slight change in fact may have produced all the mutations of climate which the researches of geologists reveal. However this may be, the facts above cited remain; they constitute true causes, the *extent* alone of the operation remaining doubtful . . . . They establish the existence of enormous differences among . . . gases and vapours as to their action upon radiant heat. . . ." Tyndall didn't use the expression "greenhouse" but he had in fact discovered that unique facet of the earth's atmosphere.

By 1896, the Swedish chemist and Nobel Laureate Svante August Arrhenius recognized that carbon dioxide allows solar radiation to pass unimpeded through the atmosphere but sequesters a portion of the energy (heat) as it is reradiated by the earth. He also calculated the effect of added carbon dioxide (carbonic acid) on the earth's temperature. Without benefit of computers, he found that a doubling of CO_2 would produce a 6°F (3.3°C) increase in mean global surface temperature (1). And writing in the *Quarterly Journal of the Royal Meteorological Society* for 1938, G. S. Callendar estimated a 2°F (1.1°C) increase in air temperature for a doubling of carbon dioxide (2). These were laborious calculations with potentially dire portents. However, as mid-century approached, the carbon dioxide theory of climate change had been all but ignored. Experiments appeared to have shown that water vapor absorbed as much of the sun's long-wave radiation as carbon dioxide. To his credit, however, Gilbert N. Plass of Johns Hopkins University took another look. His reappraisal established carbon dioxide as the major "greenhouse" gas. Plass wrote that a "relatively small change in the average temperature can have a large effect on the climate . . . . If the total amount of CO_2 in the atmosphere–ocean system is reduced by a small amount from its present value, theory predicts that the climate must fluctuate between periods with large ice sheets and warmer periods that have relatively small or no ice sheets." And he continued: There always exists a range of values for the total CO_2 amount in the atmosphere–ocean system, such that the climate continually oscillates between a glacial and interglacial stage . . . if the total CO_2 amount is reduced 50 percent or less of its present value, then a permanent period of glaciation results until the total CO_2 amount again increases." Finally, Plass contends that "the burning of fossil fuel . . . had greatly disturbed the CO_2 balance. If all this additional CO_2 remains in the atmosphere, there will be 30 percent more CO_2 in the atmosphere at the end of the twentieth century than at the beginning. Man's activities are increasing the average temperature by 1.1°C per century" (3).

Could these have been seen as a warning? The places of publication and tone of the articles do not reflect or suggest urgency. An article appeared in 1957 that with 20/20 hindsight could be said to have been a call to arms. That year, Roger Revelle and Hans Suess of the Scripps Institution of Oceanography, at La Jolla, sounded the tocsin. "Human beings," they wrote, "are now carrying out a large-scale geophysical experiment of a kind that could not have happened in the past nor be repeated in the future." It was the subsequent thought, however, that deserves—requires—full attention. "Within a few centuries," they continued, "we are returning to the atmosphere and oceans the concentrated organic carbon stored in the sedimentary rocks over hundreds of millions of years" (4). It was indeed the essential point. But who was reading; who was listening? This gem appeared in *Tellus*, an esoteric Swedish journal with a highly circumscribed readership of research meteorologists and geoclimatologists, and which few university libraries carried. Today, the Revelle–Suess admonition is widely quoted; but it took fully 20–30 years to come off the dusty shelves and gather momentum.

Nor did Soviet participation in international climate conferences and discussions raise the degree of concern. In 1971, Mikhail I. Budyko, then Director of the Voeikov Main Geophysical Observatory, Leningrad, presented his views on future climate change to The American Geophysical Union. He predicted that by the middle of the 21st century the mean air temperature could rise by several degrees. But let him speak for himself: "In the most developed industrial regions, the temperature could grow by a value of the order of one degree, and in large cities by ten degrees, a fact that would make the life in these regions impossible." As for urgency, he had this to say: "The recently recognized possibility of the change in global climate in the comparatively near future requires a closer attention to this problem than heretofore . . ." (5).

Perhaps it was the article "Carbon Dioxide and Climate: the Uncontrolled Experiment," by four scientists at the Oak Ridge National Laboratory, published in 1977 in the more widely circulated journal *American Scientist*, that not only provided momentum for the Revelle–Suess concern but also predicted that by the year 2075 if levels of atmospheric CO_2 doubled, mean global temperature would rise by 1–5°C. Again, however, urgency was limited to suggesting to readers that "the momentum of societal fuel-use patterns may make it difficult then to adjust from fossil energy to non-fossil energy quickly enough to avoid eventual severe consequences. Hence the time available for action may be quite limited." Their conclusion was loud and clear: "If the severe economic and political repercussions that are likely on a world scale are to be avoided, a technological commitment must be made in the next few years and a world strategy arrived at with enlightenment and wisdom. Though humanity may not be able to foresee the consequences of the 'great experiment' clearly enough to control them, we cannot afford not to try!'' (6). They did indeed conclude with an exclamation point, which itself must have initiated a battle between the authors and editor which they obviously won. Scientific journals scrupulously avoid emotion, let alone passion. In this instance, neither the authors nor editor can be faulted. The issue was seen to strike at the heart of civilization as we know it. Nevertheless, this article never made it to the newspapers, radio, or TV. Nor did it raise or capture the interest of *American Scientist* readers. If indeed this were a call to arms, the trumpet blast appeared to fall on deaf ears. One Letter to the Editor was all it generated. For 2 years from its date of publication, a number of articles on important but far less crucial concern for humankind drew many letters. And the lone respondent objected thoroughly to their conclusions, suggesting a panoply of other interpretations of the data and noting that ''the warming of the earth's climates may be a blessing for a hungry world. Whatever happens, we have no reason to believe that there will be no further changes in the earth's biological, geological, and physical conditions. We may as well try to enjoy the fruits of the change, whatever they may be'' (7). The author's spirited response fell flat. They appeared to have tried to convey an important message as best they knew how. Going public, to the press, radio, and TV, was simply out of the question; unthinkable for scientists. But if the issue were truly of surpassing import for mankind, who but the scientists should have alerted the leaders—the highest government authority. There was precedent for that. In 1939, Albert Einstein, Enrico Fermi, Leo Szilard, and other physicists warned President Roosevelt directly of Germanys obtaining a lead in developing the fission of uranium, an atomic bomb. The President responded by setting up the Office of Scientific Research and Development to undertake a similar project. Before the end of 1941 Fermi and others had achieved the first self-sustaining nuclear chain reaction. Urgency was there. A war was in progress. It had to be won now. Perhaps that's the key. Projections of a temperature rise by the year 2040, 2060, or 2075 may be meaningless to a here-and-now oriented society.

Nevertheless, a degree of urgency was in the making. The year 1957–1958 was declared an International Geophysical Year, and one of the projects included was the data gathering of levels of atmospheric carbon dioxide. Charles D. Keeling and Robert Bacastow of the Scripps Institution of Oceanography set up their gas analyzers near the summit of Mauna Loa on the island of Hawaii, and the South Pole station of the U.S. Antarctic Program. They found that between 1959 and 1978, atmospheric CO_2 levels had increased 6%—from 315.8 ppm to 334.6 ppm. That 19-ppm increase represented an addition of 1200 billion tons—40 gigatons—of carbon into the atmosphere in less than 20 years. Although the Mauna Loa data did in fact raise the world's climate consciousness, another decade would be required for governments to consider these data with a degree of conviction.

At the Institute for Advanced Study, Princeton, New Jersey, Jules Charney had demonstrated in 1950 the feasibility of using computers for weather prediction. By 1956, the first attempt at modeling global climate was made. But it remained for Syukuro Manabe and Richard T. Wetherald of the Geophysical Fluid Dynamics Laboratory (GFDL) at Princeton to develop the first climate model in 1967 (8). By 1980, models using the Keeling data suggested similar trends. Hansen et al. (9) and Manabe and Stouffer (10) and Manabe and Wetherald (11) estimated that doubling of atmospheric CO_2 would result in average global temperature increases of 4°C and 2°C, respectively. The differences in their estimates appeared to arise from differing assumptions about clouds and oceans. Again, these estimates remained the province of climatologists who could comprehend and grapple with them. It was not until the worldwide heatwave of the summer of 1988 and the stunning public remarks of a government scientist that urgency entered the discussions of a threat of a worldwide climatic shift.

James E. Hansen, atmospheric physicist and director of NASA's Goddard Institute for Space Studies, testifying before a Senate subcommittee, sounded the tocsin with a force and authority rarely encountered in an established scientist. The earth was getting warmer and he was 99% certain that the accumulation of greenhouse gases was responsible for the warming trend. The media embraced the story and the threat of a potentially overheating planet raced across the country and around the world. Suddenly, it moved directly from academic journals and conferences to the fast lanes of the electronic media. The underlying preoccupation with a greenhouse warming is not simply the 0.5–0.6°C increase in global temperature that has been estimated to have occurred over the past 100 years. It is rather that substantial increases in atmospheric levels of CO_2 have accumulated and persisted over the same period. If this continues, and given the current level of energy use there is no reason to suspect that it will not, additional global heating is being predicted. It is this increased warming that is worrisome as significant evidence indicates that only small changes in average surface temperature can shift our stable, hospitable climate to highly inhospitable warmer levels.

The intense concern focuses on the term "mean" or "average" as in "mean global surface temperature." Mean refers to the fact that an increase of 3, 5, or 7 degrees is an average of local temperatures from pole to pole and in-between. When temperatures from cities around the world are lumped together and an average obtained, the overall air temperature is approximately 15°C (59°F). Recall that 37°C (98.6°F) is our internal body temperature and that warm indoor swimming pools are often set at approximately 28°C (82°F) and you see that the average—over winter and summer—from cities around the world—is pleasantly cool. And in the case of the predicted 2–5°C average warming, we can be lulled into a false sense of security because the average tends to abscure both the higher and lower values that produce it. Thus, it should be clear that many cities could have far higher yearly temperatures than is suggested by the average. Furthermore, it will be useful to recall that during the last ice age 12,000–18,000 years before the present (btp), the mean temperature was no more than 5°C (8°F) colder than today. It is evident then that it does not require large temperature shifts, on average, to dislocate the world climate and thereby local weather patterns. This is a point worth digesting and storing. It is not well or widely appreciated. It should be, because it is fundamental to the problem at hand. If it were widely understood, I suspect there would be greater agitation over the possibility of a destabilizing man-made warming.

Hansen's testimony came at a propitious moment: it was hot. Weeks of drought had parched the American grain belt, and the North and Southeast were in the throes of an extended hot spell. In India the state of Rajasthan had reported 500 deaths in June when temperatures soared past 120°F for 2 weeks. The public mind was focused as never before. Hansen's message was clear; if his climate models were approximately correct, the greenhouse warming of the 1990s would be sufficient to shift the probabilities such that the chance of another extremely hot summer in most of the United States would be in the range of 55–80%. People were listening. Hansen not only had heat working for him, but 20 years of environmental consciousness raising to build on.

Nevertheless, questions were raised, as well they should be, given a subject of such grand complexity and inherent uncertainty. Tom M. L. Wigley of the Climatic Research Unit, University of East Anglia, noted that although the 1980s had the warmest years of the century, temperature levels in the higher latitudes of the northern hemisphere, where increases would have been expected, did not rise as fast as models predicted. If the decade of the 1990s is as warm or warmer, the presence of an adverse greenhouse effect would be hard to deny. And Michael E. Schlesinger of Oregon State University suggested quite reasonably that models may be more sensitive than nature. This is a point that bears remembering. Others were quick to point out uncertainties inherent in general circulation models, such as the effects of clouds and of oceans, which cover 72% of the earth's surface.

Perhaps of overriding importance was the overestimate in the predicted rise in mean sea level as a consequence of melting ice sheets. The initial belief in a 1-meter or more rise has since been reevaluated and reduced to one-third of a meter. And a doubling of CO_2 originally calculated to occur by the year 2010 is currently estimated as no sooner than 2050. Obviously, there is room to navigate among these shifting shoals. Recently, Richard Lindzen, Sloan Professor of Meteorology at MIT, maintained that "both the data and our scientific understanding do not support the present level of concern." He believes that rather than a warming the atmosphere will react to the increased level of gases with a countereffect negative feedback mechanism, which will neutralize the warming. Rather than a 1.5–4.5°C (2.7–8.1°C) anticipated temperature increase, the atmosphere would warm by little more than a few tenths of a degree. Clearly, scientific opinion is of more than one mind on this issue.

Government planners and decisionmakers could hardly be expected to move precipitously to curtail industrial output on the basis of conflicting observations. Nevertheless, given that the many calculations—extending back almost a century—were in the same direction, if not degree, greater response would not be inappropriate. However, the question that policy-makers and planners must contend with is just how large a threat the country and the world are facing. It is a question that must be asked. Not to raise it would be incompetent and derelict. But who could respond and with what degree of certainty? Would it be enough for decisionmakers to call for a reduction in current levels of energy use? What level of reduction should be sought? And by when?

The ramifications of such a decision could adversely affect every level of society. Who would be willing to throw the switch given the available data? Climate modelers can indicate trends, but their data, as good as they are, are fraught with uncertainty. The task they have set for themselves is simply to predict the degree of global climate change 50–100 years from now. This is light years more complex than predicting if rain can be expected next Thursday, or if the public will accept a new soft drink.

For climatologists to predict the effects on global temperature of doubling or trebling atmospheric carbon dioxide levels, they must know what makes climate in the first place, or what makes normal climate. That is, what set of events or conditions acting in concert has kept global climate stable over the past 18,000 years. Do they know this? If they do, they can directly program their computers to respond to injections of a range of CO_2 concentrations providing straightforward responses and the amount of temperature increase to be expected per unit of CO_2. They would then easily respond to additional questions, such as: When would a warming trend begin, if it hasn't already? Would the predicted climate change affect one area more severely, and if so, which? Is there time to institute preventive measures? This is terribly important because it dictates approaches to the problem. If there is not time, do we bestir ourselves? Does it make any difference? If there is nothing to be done, if the moment is past and we have missed the opportunity, we certainly ought at least to know it. If, on the other hand, severe drought is predicted for Nebraska, Iowa, Kansas, Missouri, and parts of Illinois, is there scientific or technologic information available that would help alleviate or minimize the problem. Or is the problem simply too large to encompass in any meaningful way? Clearly, these are not questions for climate modelers, but once their computer-generated maps indicate where adverse climate change can be anticipated, these are among the questions that must follow.

Essentially, the current problem flows from time trends; that is, two events appear to be causally related as a consequence of their rising or falling together over time. This can certainly be the case. However, time trends can also be traps for the unwary: post hoc, ergo propter hoc is one of the most common. Does the fact that both CO_2 and temperature increases appear to have occurred synchronously mean that one has caused the other? And which caused what? One of them may have caused the other, then again, perhaps not. Many events can be shown to move together in time. Lung cancer can be shown to have increased along with the use of asphalt on highways in the United States, the number of creosoted telephone poles, and the number of bananas imported, as well as the number of cigarettes smoked over the past 30 years. Are they all primary causes? Hardly. There is yet another consideration. Paleoclimatic records indicate that over geologic time our climate has been cooling, and that the earth is passing through an interglacial period on its way to an expected glacial epoch. Can an increase of CO_2 derail this schedule? What in fact do we prepare for, fire or ice?

Few of us think in terms of periodic cycles—glacials and interglacials. Periods of 10,000 or 100,000 years are totally foreign to us. After all, the average life span is measured in three score or four score and ten years. Fall follows summer and winter melts into spring, "and the voice of the turtle is heard on the land." It is all so steady and stable. Is it really? Of course, everyone knows we are living in modern times. But it may not be widely appreciated that "modern times" is synonymous with the Quarternary period of geologic time. This includes two epochs: the Holocene, which extends back 100,000 years btp; and the Pleistocene, which extends back 2 million years btp—to the origins of man. During this modern period, we have had some 10 glacial and interglacial periods, and planet Earth has moved a number of times from greenhouse to icehouse to greenhouse. Cooling and warming are natural events occurring with some regularity over these past 2 million years. And if a natural cooling trend is on the way, why is there concern for a warming trend? Can the interference of humans be discerned? Is our current problem an unnatural event? Can man change the climate?

Clearing land for agriculture and living space, prairies turned to wheat fields, wetlands to rice paddies, irrigation, dams, canals, and cloud seeding are purely human activities. The human need for food, clothing, and shelter places unusual stress on the biosphere, especially the discovery of fire with its colossal consumption of fossil fuels and subsequent release of staggering quantities of carbon dioxide. Two hundred years of the industrial revolution may have made a difference. The all but insatiable demands for wood, coal, oil, and gas have literally changed the direction and speed of atmospheric chemical evolution. Carbon dioxide is accruing in the stratosphere at a rate thousands of times faster than normal. Man may indeed be a geological force to be reckoned with, a force on a par with nature, and as such may have short-circuited the natural cooling trend and replaced it for a time with a warming period. But that remains to be seen.

The time to the next glacial epoch is variously estimated as 10,000–50,000 years, which suggests sufficient time to deal with that problem. If, however, a sudden abrupt warming interdicts the cooling trend, serious environmental dislocations may occur to our habitats around the world. On the other hand, depending on the location, unanticipated climate shifts may prove beneficial. Little has been said about this possibility—until recently. Reenter comrade Budyko. At a recent meeting in West Germany, the Soviet Union's leading climatologist, and currently Director of the Leningrad State Hydrological Institute, stated that rather than expend our efforts on holding back the warming, we should do nothing to prevent it. He believes that the warming will be beneficial and should be encouraged. Dusting off his crystal ball he sees cattle grazing in what is now the dry Sahara and crops growing in the currently arid steppes of Central Asia. He dismissed fears that the grain belts of the American Midwest and the Russian Ukraine would permanently dry out. He maintains that drying will be temporary and will give way to wetter periods and bumper crops by the middle of the 21st century provided the warming is not interdicted. "Limiting carbon fuel consumption," he stated, "will be not only useless, but even dangerous." American climatologists remain skeptical of Budyko's latest predictions.

Clearly, no amount of debate will settle the issue to anyone's satisfaction. Research may. But time may be a limiting factor. The public can rest assured, at least on this point, that the best minds in the country and around the world are literally racing for needed answers. But the issue will turn on how their data will be used by our politically oriented decisionmakers. What type of data? Most pressing is the need to know if long-term climate variations are the result of changes in the earth's astronomic orbital geometry, increases in burning of fossil fuels with subsequent release of CO_2 that can be modified, or possibly the consequence of tectonic forces deep within the earth and as such also unrelated to human activity and unmodifiable. These are fundamental questions that beg for answers. But let us be clear about one thing. It is not as if we've been caught napping, asleep at the switch, as it were. There has been, as noted earlier, a long and steady chronology behind the warming issue. It has not been sprung on an unsuspecting public; although as with most esoteric, complex, technical subjects, it could be interpreted that way. In fact, to be charitable, few have been listening. Our scientists have all but been ignored. For Hansen to go public with his data was totally uncharacteristic of the scientific persona, and he has paid a heavy professional price for his action. One can only wonder what would have occurred if Arrhenius had gone public with similar findings in 1896, or Plass in 1958. But then the media were not nearly as dramatic or prevalent as they currently are, nor were there concerned environmental groups ready to join the fray.

Currently, we have urgency. But as is characteristic of our society, instant answers are demanded of problems of planetary dimensions. We are in fact barely on the threshold of understanding the workings of the natural world. For example, we cannot yet successfully predict the weather a week in advance. Given the state of that art, can our decisionmakers be expected to discourage and limit energy production and use on the basis of computer-generated estimates, especially with the memory—still fresh—of a world food crisis predicted by a computer model in the 1970s? For many it doesn't seem to matter.

Addressing a meeting of the UN-sponsored Intergovernmental Panel on Climate Change, President Bush's message was clear. "Wherever possible," he said, "we believe that market mechanisms should be applied and that our policies must be consistent with economic growth and free market principles in all countries." If you were the head of the U.S. EPA, you were disappointed with so cautious an approach. If you were an industry representative, you praised the speech as consistent with economic growth. Representatives of environmental groups left the meeting chagrined. They had hoped for a vigorous message stressing immediate energy conservation and efficiency. Could the President have done more than straddle mine fields of conflicting views? Probably not at this time. Environmentalists are certain, and have been since 1988, that the available data are in fact a signal that the warming is already upon us—a view inconsistent with the diverse scientific data. They are also convinced that the President is vacillating—or worse, a pawn of industry. Unfortunately, the industrial attitude is business as usual until the data are certain—which may leave little margin for environmental adjustments. More than likely, general public opinion is scattered from one end of the opinion spectrum to the other, with most not knowing what or who to believe. Again unfortunately, the media have done little to disseminate objective information.

Whether a warming trend is or is not imminent, it is not a domestic issue for the United States or any other country, even though the United States is one of the major contributors of carbon dioxide and other infrared absorbing gases. This is a global issue for which little will be accomplished unless and until the nations of the world come together to deal with the problem. Nothing less than a binding treaty on energy use will suffice.

To this end, the United Nations Intergovernmental Negotiating Committee on Climate Change met in February 1991 in Chantilly, Virgina. After 10 days of often acrimonious discussion, delegates from 130 countries adopted a severely constrained pact that set out a framework for a treaty to curb the threat of global warming. In addition to the pact, it established guidelines for considering financial and technical assistance to those less developed nations that could be harmed economically by the pact. Clearly, the White House and John Sununu, its master greenhouse strategist, were able to hold off the nations of the world bent on setting specific CO_2-tonnage reductions and target dates.

The meeting in Chantilly was the first of four to be held around the world prior to the meeting scheduled for June 1992 in Rio de Janeiro. There, delegates of the nations of the world are supposed to come together to consider a treaty. Until then, the official U.S. position continues to be that it makes no sense to place a number, a level of reduction of CO_2 emissions to be achieved by a specific date, at the outset when the magnitude of the problem remains shrouded in uncertainty. It is Sununu's belief that to severely restrict the use of fossil fuels at this time would produce worldwide economic stagnation. Thus the 10-day meeting was relegated to the sole purpose of establishing guidelines for negotiating a treaty—a procedural rather than a substantive process. Accordingly, two committees or working groups have been established. One will prepare a draft treaty, which will consider commitments for limiting emissions of greenhouse gases, while the second group will deal with the legal language that a final treaty might contain if it is to effectively counter CO_2-induced climate change.

For those who see June 1992 and Rio de Janerio as the time and place of a final solution, it may be well—and quite reasonable—to consider the Chinese proverb, which informs us that if we are to know the road ahead, it would be wise to ask those coming back. In this case, those coming back could well be those who dealt with the Law of the Sea Treaty, which took 4 years to set a framework and another 9 years to negotiate, and which yet remains to be fully implemented because the United States still has not signed it. A treaty with such far-reaching and extensive controls on the energy expenditure of nations strikes at the very heart of their sovereignty. Thus, for example, it is difficult to imagine that China, with huge reserves of coal and little foreign exchange credits, would agree to a meaningful reduction on coal use just as it attempts to increase productivity and catch up to the developed world. With differing objectives, collective action may prove exceedingly elusive. Thus it is more than likely that a treaty on the air will take far more than a year to conclude. It cannot be a hasty or instant pudding. And it surely will require regular review and reconsideration in light of ongoing research and acquisition of new data. If anything is uncalled for at this time, it is a hasty conclusion.

Given the importance of the warming issue, it is sadly unfortunate that the U.S. government and many of the other major players have been preoccupied with the Gulf War, and the USSR with its own economic survival, while inflation and recession are problems for almost everyone. Not only do these divert energies but they surely color attitudes about energy use.

Global warming has captured the imagination. It contains the stuff of doomsday scenarios and thus allows the imagination to soar. It is in many ways cathartic. But all the drama and therapeutics notwithstanding, at some point, sooner better than later, the public must pick up the issues and deal directly with them. They must become involved in decisionmaking—decisions that will seriously affect them one way or another. In a complex world in which the people have become such vigorous players, it is essential that the formidable issue of potentially dislocating temperature increase be well understood. For as James Madison observed, "knowledge will forever govern ignorance: a people who mean to be their own Governors must arm themselves with the power that knowledge gives." This book seeks to provide a basis for that knowledge.

Accordingly, I have set forth five major areas which I believe encompass the issues and provide the type of information needed to comprehend the range of concerns engendered by the possibility of global warming. Therefore the first chapter provides current thinking about the reasons for our seasons. Our planet's tilt, eccentricity, and precession hold the keys to climate, but these can be slippery and elusive concepts. Perhaps readers will agree that herein they become readily available and comprehensible.

From this larger picture I have, in Chapter 2, focused on the atmosphere, which is responsible for planet Earth having the most comfortable climate to be found anywhere in our solar system. It is of course this most humanly comfortable of climates, a mild 59°C, that has allowed humankind to thrive these past 10,000 years. And it is the atmosphere that appears to be threatened by the gaseous by-products of human activity as we search for greater comfort and nourishment for ever larger numbers of people, many of whom are tired of animal-like drudgery and a subsistence existence. They want more. They need more. They should have more. But "more" raises a perplexing situation. The threat of upsetting the climate system, which has so remarkably sustained mankind's development thus far, could well demand a damping of the fires as it were—the curtailment of burning fossil fuels, which would seriously impede development. Thus Chapter 2 discusses the natural gaseous balance as well as defines and describes what has popularly come to be known as "the greenhouse" and the greenhouse effect. In addition to the natural constituent gases, the atmosphere now contains a range of chlorofluorocarbons (CFCs), manufactured gases that contribute mightily to two distinctly different untoward atmospheric problems. The CFCs are treated in detail.

In Chapter 3, I move to a discussion of models and their use in predicting global warming and its by-products—the consequences that could flow from an increase in global temperature. Here the predictions of a warmer world are discussed in terms of the elements involved in developing forecasts, the use of models and their complexity, as well as the uncertainty inherent in such dicey business. While prophecy is always risky, gamblers, unlike scientists, thrive on it. But the outcome of a horse race or a football game, launching another breakfast cereal or mouthwash, or buying pork bellies for delivery next March bear little resemblance to the dire consequences for us all should decisionmaking based on computer-generated models be wrong—or we fail to heed their warning. Are we hoist on our own petard?

Chapter 4 reaches into the broad range of potential ecologic dislocations predicted by models, to confront such momentous issues as sea level rise, hurricanes and storms, and threats to agriculture and human health. Discussed too are the benefits to our food supply that may accrue. Increased CO_2 in the air may well stimulate higher per acre yields of a variety of plant crops. This is a vital element for the total equation. And this chapter discusses the potential adverse health effects of increased contact with ultra-violet light as a possible consequence of an ozone-reduced atmosphere. But it also offers opportunities for prevention of UV- induced cell damage. The chapter concludes by asking what we know for certain, and what remains uncertain. The answers may be surprising.

The final chapter offers realistic solutions to the energy problem, solutions that relatively easily can reduce energy consumption and much waste, which cheap energy has wrought. In addition, Chapter 5 confronts the serious problem of the general and widespread lack of knowledge, interest, and concern with science and technology, as well as the need to modify behavior in an energy complacent society.

While I have tried to present an overview of this most serious environmental issue, providing readers with in-depth information and far greater scope than can be gained from the limited perspective of computer models, which unfortunately has usurped media discussions, my personal biases may be evident. How successful have been my efforts to confront and extirpate them remains for each reader to decide.

There is room for argument and differences of opinion in this most serious developing problem—perhaps the most serious humankind has yet encountered. The answers are by no means in—a vocal few notwithstanding. The critical decisions that must be made for this country as well as the international community call for a well informed citizenry. I would like to think that this book would move many in that direction.

REFERENCES

1. Arrhenius, S. A. On the Influence of Carbonic Acid in the Air upon the Temperature of the Ground. *Philos. Mag.* 41 (5):237–276, 1896.
2. Callender, G. S. The Artificial Production of Carbon Dioxide and Its Influence on Temperature. *Q. J. R. Meteorol. Soc.* 64:223–240, 1938.
3. Plass, G. N. The Carbon Dioxide Theory of Climatic Change. *Tellus* 8:140–153, 1956.
4. Revelle, R., and Suess, H. Carbon Dioxide Exchange Between the Atmosphere and Ocean, and the Question of an Increase in Atmospheric CO_2 During the Past Decades. *Tellus* 9:18–27, 1957.
5. Budyko, M. I. The Future Climate. *EOS Trans. Am. Geophys. Union* 53:868–874, 1972.
6. Baes, C. F. Jr., Goeller, H. E., Olson, J. S., and Rotty, R. M. Carbon Dioxide and Climate: The Uncontrolled Experiment. *Am. Sci.* 65:310–320, 1977.

7. Alker, J. A Future Heat Wave. *Am. Sci.* 65:532–533, 1977

8. Manabe, S., and Wetherald, R. T. Thermal Equilibrium of the Atmosphere with a Given Distribution of Relative Humidity. J. Atmos. Sci. 24:241–259, 1967.

9. Hansen, J. E., Johnson, D., Lacis, A., Lebedeff, S., Lee, P., Rind, D., and Russel, G. Climatic Impact of Increasing Atmospheric Carbon Dioxide. *Science* 213:957–966, 1981.

10. Manabe, S., and Stouffer, R. J. Sensitivity of a Global Model to an Increase of CO_2 concentration in the Atmosphere. *J. Geophys. Res.* 85(C10):5529–5554, 1980.

11. Manabe, S., and Wetherald, R. T. On the Distribution of Climate Change Resulting from an Increase in CO_2-Content of the Atmosphere. *J. Atmos. Sci.* 37:99–118, 1980.

CHAPTER 1

REASONS FOR SEASONS

$\mathbf{T}$he misery and suffering of the Continental Army at Valley Forge are legendary. For the rag-tag militia encamped in the open during the dreadful winter of 1777–1778, it was a grim ordeal. But the coldest winter was yet to come.

At his headquarters in Morristown, New Jersey, General Washington kept a weather diary in which he noted on January 6, 1780, that "the whole of the country lay buried from three to five feet deep . . . in places the drifts were piled ten to twelve feet high." These were not ordinary winters. In fact, January 1780, as David Ludlum recounts in *Early American Winters*, "was probably the coldest calendar month in recorded American meteorological history." And from primary sources therein we learn that "on January 1, so much ice filled New York harbor that vessels could not venture through the Bay and reach the sea. The continuance of ice through January eventually shut up every seaport along the Atlantic coast, even those not ordinarily susceptible to ice blockade" (1).

In Boston, "the harbor froze over so that people could pass to Castle Island and to Charlestown with loaded sleds." While at Philadelphia, "the river was fuller of ice than has been known at this time of year, for many years past, so that no vessel can come in or out till a favorable opportunity offers." The severe cold also took its toll at sea. "A ship arriving at Newport, Rhode Island, from Barbados, had several men with frozen extremities, and another brought ashore a sailor who had succumbed to frostbite when approaching the port." This exceptional cold was not a local or "New World" phenomenon. From the 15th to the mid-19th centuries, the entire northern hemisphere was in the grip of the Little Ice Age. In Europe, Hans Brinker fashioned skates to glide along frozen Dutch canals and, as shown in Figure 1, Londoners frolicked on a frozen Thames. Norwegian, Swiss, and Icelandic farmers saw glaciers advancing over their fields, while those English farmers who had planted their fields with vines and become exporters of wine now lost all to the severe cold.

Nothing remotely reminiscent of the unseasonably cold 18th century winters has occurred over the past 150 years. But cold weather has contin-

Figure 1. This 17th century painting shows the Thames as an almost solid block of ice. Ice skating was relatively new and only a few people, as shown here, had the new skates.

ued, especially at the poles. On January 23, 1971, −80°F was recorded at Prospect Creek, Alaska, and on July 21, 1983, −128.6°F was documented at the Vostok Station in Soviet Antarctica. On the other hand, it's been devilishly hot at times. On July 10, 1953, Death Valley, California lay broiling under 134 scorching Fahrenheit degrees (43°C): not the hottest temperature the world has ever recorded, but close. A record high of 136°F had occurred in North Africa, in Azizca, Libya, on September 13, 1922. A single hot day is one thing, but constant heat is another. In the summer of 1936, the citizens of Iowa suffered through temperatures of 100°F or better for 53 days. In July temperatures rose to 120°F, and 121°F was recorded in Kansas and North Dakota. And the summer of 1980 will not soon be forgotten in Dallas, Texas. From June 23 on, these Texans sweltered for 42 consecutive days of 100°F heat or higher. We human beings cannot live in this degree of heat for extended periods without serious consequences. And the sweltering 1980s gained climatological preeminence with six of the ten warmest years of the century. Although most people would prefer 1988s blistering June/July days to slip evanescently out of memory, few will forget. Nevertheless, over the past 10,000–12,000 years, mean global climate at 15°C has been hospitably mild—and stable. Predictable. Extremes of temperature notwithstanding, we live by well defined and faithful seasons— for good reason.

Within the vastness of solar space, nine planets tract their unique paths around an incandescent sun. Of the nine, the earth alone has affable seasons. Tilt, eccentricity, and precession are the reasons. The global earth is not fixed in space. It wobbles, spins, nods, rotates, and inclines. This diversity of earthly planetary motion, its orbital geometry, alters its position, motion, and location relative to the sun, thereby varying the amount of sunlight striking the earth . . . from season to season. Thus, together, the tilt of the earth's axis, the eccentricity of its orbit, and the precession of the equinoxes can intensify or weaken seasonality.

The earth is tilted. It leans substantially, relative to the ecliptic—the plane of its orbit. How great is the earth's axial tilt? The traditional response would be 23.5°—which would be correct . . . for a time. From the perspective of a single lifetime or even many generations the 23.5° may seem fixed or immutable; in fact, it isn't. Over a 41,000-year cycle the earth's angle of inclination varies some 2.6°, from 21.8° to 24.43° from perpendicular (Figure 2). Astronomers refer to this tilt as the obliquity of the ecliptic. The greater the obliquity, the more pronounced is the difference between winter and summer. If the tilt were zero, the entire globe would have 12 hours of daylight and 12 hours of darkness, along with a loss of seasons. If the tilt were greater than 25°, say, 40° or more, winters would be extremely cold and summers extremely hot. Thus the current 23.5° of obliquity is not only hospitable to life but appropriate to engender four distinct seasons.

As the inclination approaches 24.4°, winters become a little colder and summers a little warmer. When the north pole of the axis is tilted toward the sun as shown in Figure 3a, summer comes to the northern hemisphere. New York, Baltimore, and Washington, for example, are closer to the sun.

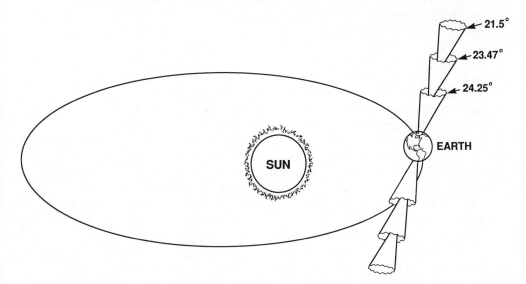

Figure 2. Over a period of some 41,000 years, the earth's angle of inclination varies from a low of 21.5° to 24.25° from perpendicular. Note too that the orbit is not smooth but nutated: that is, with the earth's oscillation about its poles, the orbit takes on a wobbling quality.

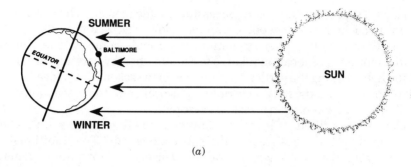

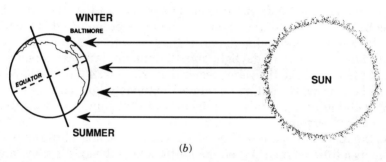

Figure 3. The tilt of the earth's axis as it revolves around the sun is a major reason for our seasons. (a)When the north pole is inclined toward the sun, the sun's rays strike the earth more directly and are more concentrated, and summer comes to the northern hemisphere. (b) In winter in the northern hemisphere, the north pole is inclined away from the sun. The days are shorter and colder as the sun's rays become less concentrated. The reverse is true in the southern hemisphere.

In the southern hemisphere the seasons are reversed. It is winter in Rio de Janeiro, Buenos Aires, and Santiago. When the north pole is inclined away from the sun, Figure 3b, winter comes to North America and summer to the southern hemisphere. That is, because of the earth's tilt, the northern hemisphere receives more solar radiation for half of the year, and the southern hemisphere more for the other half.

During winter in either hemisphere, the sun strikes the earth at oblique angles, which do not concentrate heat as much as they do when striking vertically—at right angles. As shown in Figure 4, a beam of light striking the earth at right angles, or close to it, provides twice as much heat as a ray hitting at an oblique angle. An increase in tilt from 21° to 24° concentrates radiation in the upper latitudes (above 65°N) and the poles, while reducing incoming solar radiation—insolation—at the lower latitudes. Currently, the tilt is decreasing, which suggests that differences between winter and summer are less today than they were some thousands of years ago. Thus

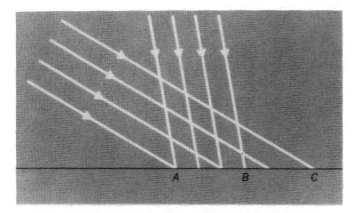

Figure 4. Rays of light striking the earth directly do so with energy concentrated over a narrower area. Light rays striking at oblique angles cover a wider area and thereby cast weaker beams.

summers now are supposed to be a little cooler and winters a little warmer. That's a point worth recalling.

Tilt seasonality does not operate alone. Seasonality is also affected by the shape—eccentricity—of the earth's orbit. Thus to the question "Is the path traversed by the earth around the sun circular or elliptical?" an anwer must be framed in terms of "when?" In fact, the orbit as shown in Figure 5 changes from nearly circular to somewhat elliptical over a period of 100,000 years. Currently, the orbital path is less than 2% off round—barely elliptical. Over the past 100,000 years the degree of eccentricity has varied from near zero to about 6%: actually 0.017 (1.7%) to 0.054 (5.4%). Using Figure 6 for comparison of intensity of eccentricity, up to 6% must be seen as rather weakly elliptical—far less than the tightly stretched, elongated ellipses often suggested in artists' renderings. Nevertheless, as weak or as

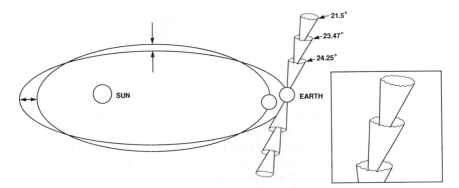

Figure 5. The earth's orbital path around the sun shifts from nearly circular to elliptical every 100,000 years.

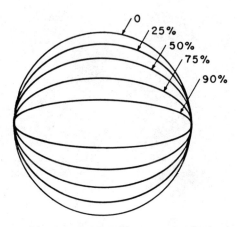

Figure 6. Ellipses with eccentricities of from zero (perfectly circular) to 90%.

small as they are, the different eccentricities represent large differences in earth–sun distances and as such further affect the intensity of solar radiation reaching the earth. Figure 7 shows the type of change in earth–sun distance that orbital shifting can produce. When winter in the northern hemisphere occurs with the earth closer to the sun, as it currently does, winter can be expected to be milder and summers warmer. This is an important consideration. Given this configuration, summers would be expected to be warmer naturally. Climatologists refer to this type of climate change as orbital forcing—forcing being a change that modifies the earth's radiation balance, its insolation, and thereby its temperature.

Although the spinning earth traverses the 600-million-mile orbital path in 365.25 days, it must be recalled that the orbital cycle of more nearly circular to elliptical to circular requires approximately 100,000 years. These long periodicities, 41,000 and 100,000 years, are pivotal to an understanding of past, present, and future climate. For its effect on seasonality, however, the crucial elements in orbital eccentricity are the periodic increases and decreases in sun–earth distances. As distance increases, seasons become milder in one hemisphere and more intense in the other. At perihelion, the point on the orbit at which the earth is closest to the sun, the earth receives some 3.5% more solar radiation, while at aphelion, the point of the orbit at which the earth is farthest from the sun, it receives some 3.5% less. On a more circular orbit the earth receives about the same amount of heat daily. As orbital stretching occurs, heating becomes uneven, but the total amount of radiation (heat) received during the year remains constant. Eccentricity and tilt affect only the seasonal distribution. These differences are most evident in the temperate zones and least at the equator. Thus tilt, the obliquity of the ecliptic, plays a major role in seasonality, and orbital eccentricity intensifies or weakens the effects si-

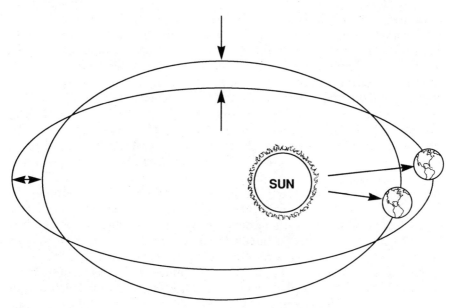

Figure 7. Over a period of 100,000 years, the earth's orbital path shifts from nearly circular to elliptical. Consequently, the earth–sun distance must also change.

multaneously. Consequently, changes in the distribution of sunlight—the heat balance—are the major reasons for our distinctive seasons.

Although perihelion and aphelion are the points along the orbit at which the earth is closest and farthest from the sun, these are not fixed points. They might be if the earth's orbit were a perfect circle. In that case, each season would occur at the same distance from the sun. However, given the tug of gravitational forces exerted on the earth by the sun, moon, and planets, the orbit is less than circular. And a third orbital force, precession, determines whether the effect of tilt is enhanced or reduced by orbital eccentricity. Thus it is precession that determines whether a season in one hemisphere falls at or near a far point along the orbit. Precession is thus the third element of the triad, which together constitute the mechanism of natural climate forcing. And it is precession that prevents fixing of the points of perihelion and aphelion.

Precession has two components, axial and elliptical, which operate simultaneously to effect a shift in the four equinoctial points along the orbit. The shift in these points, a little earlier each year—hence the term precession—is the precession of the equinoxes. These points mark the beginnings of each of our four seasons: December 21, winter and with it the shortest day of the year, sets the winter solstice; March 20 and September 22, the vernal and autumnal equinoxes, are the two dates when day and night are of equal length; and June 21, summer and the longest day of the year, is the summer solstice.

Consider the earth as a huge toplike body, which like a spinning top wobbles as it spins . . . on its axis. The faster the spin the less the wobble, the slower the spin the greater the wobble. As a consequence of its spinning and wobbling as well as the effects of orbital eccentricity, the earth precesses or gyrates about its axis. As shown in Figure 8, as the earth tilts or tips in one direction its rotating motion reorients the spinning globe, setting in motion a new tilting process. The net result of this continuing process of tilting and reorientation is that the axis of the wobbling earth precesses, sweeping out an imaginary cone in space. That is, if the north and south poles extended above and below the earth, and each were tipped with a phosphorescent material, the tips, seen from above and below, would appear to move slowly westward in a circular motion through the zodiacal stars, and the width of the cone, as shown in Figure 9, would vary as the angle of obliquity varied from 21°to 25°. A complete rotation requires some 23,000 years. Thus we have a third long periodicity that affects our seasons and our lives. This rotation shifts direction such that "the pole star"—north—changes over time. Today, and for the past 2000 years, navigators have been guided by Polaris, at the end of the Little Dipper. In 3000 B.C., the north pole projected a line toward the star Alpha Draconis. At about that time the Great Pyramid built for Khufu at Gizeh was completed and Egyptian astronomers aligned it with that north star. Today the Pyramid no longer points north. However, it will again as axial precession brings the celestial north pole once again to that point on the sky. In A.D.

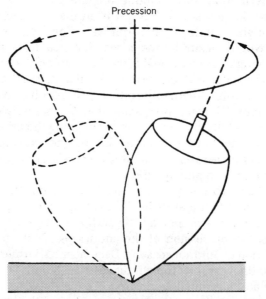

Figure 8. Precession can be compared to the clockwise spin of a whirling top whose spin vector traces out an ellipse as it sways from side to side.

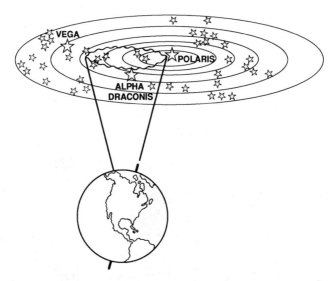

Figure 9. During its 23,000-year precession cycle, the earth's rotation points the pole toward different sites in the sky. Accordingly, in 3000 B.C. Alpha Draconis was the north star. Currently, Polaris is the north star, and in A.D. 14,000, Vega, the brightest star in the heavens, will be the north star.

5000, if sailors are still looking to the heavens for navigational guidance, they will be looking for the star Alpha Cephai, and in A.D. 13,000, Vega, the brightest star in the sky, will become the pole star. When, 23,000 years from now, this precession cycle is completed, Polaris will once again become the north star.

While the earth wobbles, the orbit is rotating independently and as it does it shifts the equinoctial points 50 seconds of arc each year. With 360° in a circle, and with 1° equal to 60 minutes of arc, and 1 minute equal to 60 seconds of arc, a shift of 50 seconds alters an equinoctial point less than one-sixtieth of a degree (0.0164°) per year. At this rate it would take some 22,000–23,000 years for the equinoxes and solstices to shift completely around the orbit. And they occur slightly earlier each year—hence again the term precession.

Figure 10 demonstrates this periodic positional shift. Currently, in the northern hemisphere winter occurs at a point close to the sun. Consequently, winters will be milder and summers warmer—naturally. Six months later in June, the earth is at aphelion, its farthest point from the sun. Thus cooler summers should be anticipated. It was not always this way. "Always" is a uniquely relative term. In terms of human life spans it has one meaning, in geological time quite another. Thus 11,000 years ago—Figure 10, at the end of the last ice age, half a precessional cycle— insolation was greatest in summer in the northern hemisphere. And 11,000 years from now, elliptical precession will again bring the earth—and the

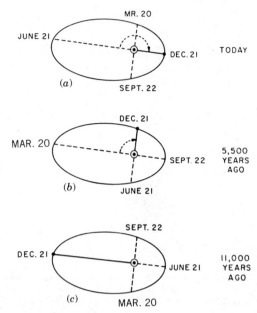

Figure 10. Precession of the equinoxes. Due to axial precession the positions of the equinoxes—March 20 and September 22—and solstices—June 21 and December 21—shift slowly around the earth's elliptical orbit and complete one full cycle about every 22,000 years. Eleven thousand years ago (c) the winter solstice occurred near one end of the orbit. Currently, (a), the winter solstice occurs near the opposite end of the orbit. Consequently, the earth–sun distance measured on December 21 changes. (Reproduced with permission of Enslow Publishing Company, Hillside NJ.)

winter solstice—to the point of aphelion, giving the northern hemisphere cooler winters and warmer summers. Precession then can determine if the seasonal effects of obliquity are enhanced or weakened by distance seasonality. When they reinforce one another a season will become intense in one hemisphere and weakened in another. Accordingly, if solar radiation is constant, variations in geographical distribution of insolation must result from the triadic changes in orbital geometry. Variable insolation, especially in the high northern latitudes, was more than likely sufficient to cause the shifts from glacial to interglacial epochs. Can this be verified? Is there supporting evidence for any of this?

In 1930, Milutin Milankovitch, a Serbian (Yugoslavian) engineer[*] and professor of mechanics and theoretical physics at the University of Belgrade, published his calculations of the intensity of incoming solar radia-

---

[*] Milankovitch has been referred to as a physicist, mathematician, and astronomer, as well as variations of these, such as mathematical physicist. In fact, he graduated from the Vienna Technological Institute with a doctorate in engineering and worked for years as a civil engineer developing new processes for reinforcing concrete. He was born in 1879 in Dalj, Slavonia, part of the Austro-Hungarian empire, and died in 1958 at age 79.

tion and stated that only small changes in insolation would be required to shift our relatively mild climate to an ice age (2). And in 1938 in a second publication, he maintained that ice ages were due to variations in radiation primarily at latitudes above 65°N and that this could ultimately be shown to be the consequence of tilt, eccentricity, and precession with their periodicities of 41,000, 100,000, and 23,000 years (3). This was a theoretical and scientific tour de force. But for most climatologists and geologists it was and remained little more than an elegant exercise. Why would anyone believe it? How could anyone believe it? These were indeed precise measurements but they could not be proved. There was no evidence with which to back it up—to even test it. And so it remained for over 30 years. Meanwhile, geologists were stitching together evidence showing that ice ages had in fact not only occurred frequently but regularly. Over the past 1 million years, as shown in Figure 11, ten such periods had come and gone, and each had lasted approximately 100,000 years. Between each glacial was a mild interglacial lasting 10,000–20,000 years. Many climatologists believe we are currently approaching the end of an interglacial and should be moving into a cooler period prior to the onset of the next scheduled ice age. Without unnatural intervention much of North America and Europe should once again be covered with ice.

That climate variability over geologic time is both natural and continuous is seen in Figure 12, the work of L. A. Frakes, Department of Earth Sciences, Monash University, Australia (4). The curves represent pos-

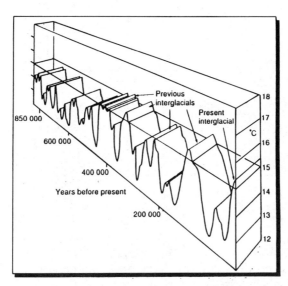

Figure 11. Over the past 1 million years the climate of the earth has shifted from glacial to interglacial periods, from cool to warm, at least ten times—naturally. Thus climate change is the rule, not the exception.

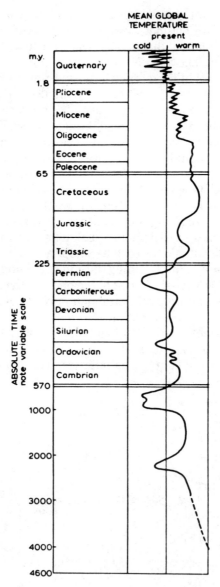

Figure 12. This generalized temperature history of the earth, extending back some 4 billion years, attests to the fact that variation is natural. The shifts here represent periods cooler or warmer than the present. (Reproduced with permission of Elsevier Publishing Company, Amsterdam.)

tulated departures from current global mean temperature, and the dashed portions indicate sparse data. Because of the abundance of paleoclimatic data for more recent times, the scale is expanded toward the top as the Pleistocene is approached. Overall, it is apparent that warm and cold periods have followed one another over hundreds of millions of years. Evident, too, from the middle of the Cenozoic, the Miocene, temperatures began a long decline that has continued ever since, warm intervals notwithstanding. As Figure 11 illustrates, over the past million years glacial and interglacial intervals have succeeded one another with rhythmic regularity. And only for this last 10,000–12,000-year interval, the period geologists refer to as the Holocene, has there been sufficient warmth for civilization as we know it to thrive. In terms of human life 10,000 years is almost unimaginable, but in geologic terms it is but a blip, given the 4.3 billion years since the earth was born. Considering that we are dealing with planetary forces, it may be necessary to begin to think in geologic terms if we are to comprehend both the short- and long-term consequences of human activity. This raises the salient point that the difference in temperature between glacial periods and the current warm interglacial is approximately 5°C (9°F). Thus the effects on insolation of changes in the earth's orbital geometry are indeed small. The sobering fact is that it may take only a change of a few degrees to shift the radiation balance. Is mankind capable of that?

As climatologists argued over Milankovitch and how his theory might be tested, and while geologists studied the earth's strata, Cesare Emiliani, a graduate student at the University of Chicago's Department of Geology, and currently a member of the faculty at the University of Miami, published a remarkable and seminal paper. In 1955, he had in fact found the key to Pleistocene climate. It would literally revolutionize the study of paleoclimatology. Here was an isotopic technique with which to unlock not past time periods but past temperature, and which also related it to orbital forcing. But in Chicago, Emiliani had excellent guidance. In 1948, for example, Harold Urey enunciated an idea that guided generations of investigators. "If an animal deposits calcium carbonate in equilibrium with the water in which it lies," he remarked, "and the shell sinks to the bottom of the sea, . . . it is only necessary to determine the ratio of the isotopes of oxygen in the shell today in order to know the temperature at which the animal lived" (5). Theoretically, here was a ready-made, in-place thermometer with which to plumb the earth's temperature as far back as shells existed. It remained only to be read. Emiliani discovered the means of reading the "thermometer."

The most well known and widely used application of isotopes is as a dating tool. A wide variety of methods based on decay or buildup of radioactive isotopes allows age determination of a broad range of materials—wood, ice, rock, and mud. By far the most popular dating method is based on the decay of the carbon isotope, carbon-14, with its half-life of 5730 years (see Chapters 2 and 3 for a detailed discussion of its use). The stable isotopes most often used in climatological studies are the heavy isotopes of hydrogen and oxygen, that is, deuterium and oxygen-18, which permit the tracking of climate-induced changes. Because of its relative homogeneity, the ocean can be an ideal environment to record long- term climatic variation. Due to its mass, short-term climatic fluctuations have a negligible impact on it. The oxygen isotope ratio of the carbonate shells of marine Foraminifera is a function of its ratio in seawater as well as the temperature of the water. Thus isotopic measurement goes beyond dating to estimation of temperature as well as volume of ice. As water evaporates from the oceans and accumulates in the growing ice sheets, the oceans become increasingly enriched in the heavier ^{18}O isotope. Hence the variation can be used to derive the environmental conditions existing at the time of uptake. Together, Emiliani and Urey theorized that the key to unlocking paleoclimatic temperatures was to be found in marine invertebrates, especially those microfauna whose bodies were housed in calcium carbonate shells or tests. But it was the oxygen atoms within the carbonate fraction that excited them. Oxygen exists in three natural forms: heavy oxygen, ^{18}O, occurring at approximately 0.2% of the total; normal oxygen, ^{16}O, at 99.76% of the total; and ^{17}O, a rare form that can be discounted. In the process of respiration, the microfauna take up carbon dioxide and oxygen, and in the process of shell development convert them to carbonate, which contains the oxygen isotopes in a ratio reflecting their relative abundance at a specific time in their aqueous environment. When the organism dies and sinks to the bottom, the isotope ratio is preserved forever in the fossilized shells as sediment covers it over and successive generations fall on top, and on and on.

Emiliani understood that the amount of ^{18}O incorporated into shells was temperature dependent: the colder the water, the greater the number of ^{18}O atoms sequestered. Therefore by determining the ratio of ^{18}O to ^{16}O in fossilized shells contained in sediments at various depths, a temperature gradient could be constructed. The formula used was

$$\delta\,^{18}O = \frac{^{18}O/^{16}O \text{ sample} - {}^{18}O/^{16}O \text{ standard}}{^{18}O/^{16}O \text{ standard}}$$

The standard used in his laboratory was a pulverized specimen of *Belemnitella americana*, a Foraminiferan obtained from South Carolina's Upper Cretaceous Peedee formation. And he determined the difference between the two by "alternatively passing the sample and the standard through a mass spectrophotometer." Negative values represent lower amounts of ^{18}O while positive values represent greater levels of ^{16}O to ^{18}O. In this way he was able to estimate the temperature of the seas over a period of 280,000 years before the present (btp) (6). Thus the sediments become a barometer, or more precisely a planetary thermometer, of bygone epochs.

Testable material for evidence of the correctness of the Milankovitch mechanism was at hand. Emiliani opened a window of opportunity through which others fairly poured.

It has now been shown that this ratio agrees well with the variation of insolation induced by the cyclic changes of the earth's orbital parameters—tilt, obliquity, and precision—the variations in orbital geometry which in the words of Hays, Imbrie, and Shackleton become "the pacemakers of the ice ages" (7). The periodic growth and retreat of the ice sheets as propounded by Milankovitch have been satisfactorily explained. Thus the earth's orbital changes alter the latitudinal and seasonal distribution of incoming solar radiation—insolation—and the global volume of glacial ice "paces the variation of insolation," with a lag of a few thousand years—the time required for the ice sheets to build and retreat. Furthermore, this variation in insolation affects climate worldwide—naturally.

The ice sheets of Greenland and Antarctica have proved to be excellent repositories—archives—of ancient (Pleistocene) climate. The vertical stratification of the snow and ice has been reasonably well preserved, although the presence of continual compaction has produced a slight decrease in the thickness of the annual layers with depth. Therefore the counting of the isotopic cycles along a vertical profile has now been shown to be an appropriate method of dating an ice core (Figure 13a), as well as indicating the rate of its accumulation. In addition, the analysis of air bubbles trapped in the ice (Figure 13b) permits the determination of carbon dioxide and methane locked in the ice at the time of formation. Thus profiles of these gases back over a continuous period of hundreds of thousands of years are now available, and they parallel the temperature curves derived from the heavy oxygen and hydrogen profiles.

For example, teams of Soviet and French glaciologists drilled into the ice at the Russian Vostok station in East Antarctica (Figure 14), and obtained ice cores down to 2200 meters (6600 feet) whose age extended back 160,000 years btp. Using the oxygen isotope technique, they were able to sample the air contained in the bubbles trapped in the ice (Figure 13). From their ice core data they constructed the most complete model of recent climate shifts. Their data showed that the rhythms of waxing and waning of the ice sheets had followed quite precisely the 100,000-year orbital periodicity (8). The picture was coming into sharper focus and strong proof was emerging that climatic changes were indeed the consequence of insolation changes resulting from periodic variations in the earth's orbital parameters.

Researchers from the University of Berne took another approach and with it opened yet another window of understanding. Rather than sample oxygen in the trapped bubbles, Oeschger and co-workers chose to investigate not temperature but the atmospheric record. To do this, they would test for the presence of carbon dioxide. As noted, ice cores contain bubbles of gas (air) that were trapped as snow fell and solidified into ice. Falling snow is light and fluffy, providing air spaces between the flakes. As snow continues to fall and the years pass, each succeeding layer presses down on the layer below, trapping the air, which closes and seals itself into a bubble

(a)

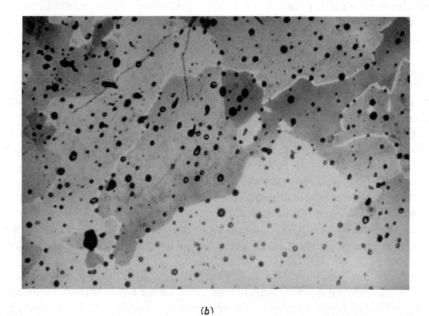

(b)

Figure 13. (a) A typical length of ice core, while relatively heavy, can be held in the palm of the hands and has great clarity. (Courtesy of Dr. Bruce Koci, Polar Ice Coring Office, University of Alaska, Fairbanks.) (b) Bubbles of air trapped in Antarctic ice. (Courtesy of Claude Lorius, Laboratory of Glaciology and Geophysics of the Environment, St. Martin d'Heres, France.) (See color section.)

Figure 14. The Soviet Antarctic Vostok Station. The photographs were taken May 1978. An interior plateau station was built in 1957 by the Second Soviet Antarctic Expedition. New quarters built in 1974 were erected on open lattice supports to prevent the accumulation of drifts. (Courtesy of the National Science Foundation, Division of Polar Programs, Washington, DC.)

over a period of 50–100 years. The placid snow is frozen solid and the bubbles of gas are preserved layer upon layer with the oldest bubbles further down the ice sheet. Centimeter-sized cubes are cut from the long cores and crushed in a vacuum chamber. The total pressure of the released air is measured and the carbon dioxide (CO_2) level determined. Oeschger and co-workers found that some 18,000–20,000 years ago, toward the end of the last ice age, CO_2 attained levels of 180–200 ppm—0.018–0.020%. As they continued to sample they found that, as the end of the ice age was reached, the level of CO_2 increased to 270–280 ppm—similar to levels of the late 18th century prior to the dawn of the industrial revolution (9). Compared to the current level of approximately 350 ppm, their data were startling in their potential implications. But were the data believable? Could they be trusted? The technique was exceedingly demanding. Figure 15 relates temperature and CO_2 over the past 160,000 years. Whether temperature-induced atmospheric changes in CO_2 or CO_2 levels triggered temperature shifts remains to be determined and could hold the key to future climate trends. The data need verification.

As if to dot the i's and cross the t's, Nicholas J. Shackleton of Cambridge University and Niklas G. Pisias of the Lamont–Doherty Geological Observatory, Palisades, New York, analyzed a 340,000-year-old sediment core containing three complete glacial–interglacial cycles for its record of both oxygen and carbon dioxide isotopes and found that the periodicities characteristic of changes in the earth's orbital geometry documented all the

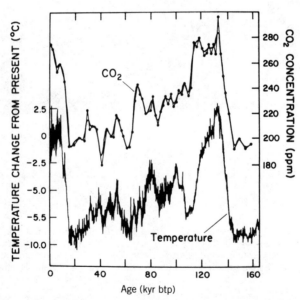

Figure 15. That there is a relationship between CO_2 and global temperature is evident. What that relationship is is not so evident. Nevertheless, these are the data that appeared to solidify the idea of CO_2-induced global warming.

records. The conclusion appears to be that major pacemaker of climate is the 100,000-year cycle. Six intervals of rapid deglaciation are indicated in Figure 16. Periods of glacial expansion averaging 100,000 years were abruptly terminated by rapid deglaciations. In Figure 16 the changes in oxygen isotope ratio in a deep-sea sediment core show a clear 100,000-year cycle. The saw-toothed nature of the curve with its six intervals of rapid deglaciation suggests that the retreat of the ice sheets occurred much faster than their advance. Shackleton and Pisias also found that atmospheric CO_2 levels varied in response to these orbital forcings and that changes in obliquity of the earth's rotational axis played the major role, and it was clear that atmospheric CO_2 in conjunction with variations in insolation appeared to control the growth and retreat of the ice sheets (10). Their work

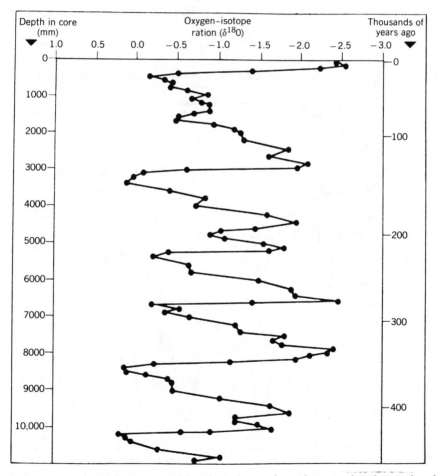

Figure 16. The saw-toothed nature of the plot indicates that at least six glacial–interglacial episodes occurred, and that the warmer interglacials occurred more rapidly than the cooler glacials.

supported Oeschger's ice core data. With firmer data now available, it was
clear that Milankovitch was correct in that orbital factors played a pivotal
role in global climate. Without their influence there would be no shifting
from glacial to interglacial periods. Thus climate change is the rule, not the
exception. It occurs regularly and it is a natural phenomenon. These are
key points. Radiation balance is a significant factor but by itself does not
appear adequate to shift the balance.

That tilt, obliquity, and precession vary naturally over time is well
delineated in Figure 17. Also evident is the distinctive alteration in amplitude of each of their periodicities. This again attests to the fact that changing patterns of insolation have been operating for at least 1 million years.
Nevertheless, orbital forcing by itself seems unable to fully explain climate
change. Another ingredient may be required.

These features suggest a fundamental link between the climate system
and carbon dioxide in addition to variable insolation, which may account
for the observed temperature history. In fact, Genthon and co-workers of
the combined French and Soviet team believe that "climate changes would
be triggered by insolation changes with the relatively weak orbital forcing

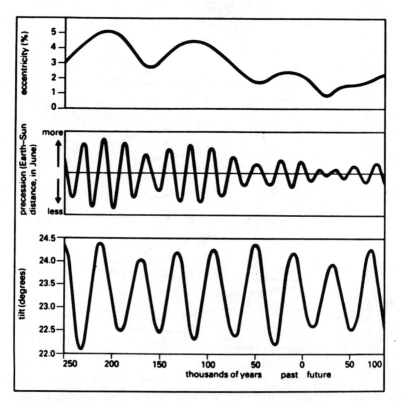

Figure 17. Changes in eccentricity, tilt, and precession. Planetary movements induce alterations in the gravitational field, which in turn alter the geometry of the earth's orbit.

being strongly amplified by possibly orbitally induced CO_2 changes" (11). However, they add a caveat that must not be overlooked: "One must realize," they inform us, "that long-term feedbacks operate quite differently when going towards the last or the present interglacial than when going towards a possible future CO_2 super interglacial." In a more recent observation, this group mentioned that "the results obtained from the Vostok ice core are convincing indications that there is an interactive link between orbital forcing, CO_2 and climate, and we support 'the suggestion' that further progress will require that climate and the carbon cycle be treated as part of the same global system rather than as separate entities" (12). The curves in Figure 11 cover approximately the past 1 million years—back to the middle of the Pleistocene—and thus include the ten or more cycles of glacial–interglacial periods, which is impressive corroboration for the constancy and regularity of the phenomenon of changing seasonality. Recently, geologist at Columbia University's Lamont–Doherty Geological Observatory obtained deep core samples that will provide a continuous 25-million-year record of cyclical climate change. As shown in Figures 18 and 19 core samples were obtained of buried ancient lake sediments (under the Rutgers University Kilmer campus, New Brunswick, NJ) with distinctive banded patterns that chronicle the earth's climate changes in the finest detail yet seen. In addition to the Milankovitch cycles,

Figure 18. Amoco workers operating a high-speed drill remove a 20-foot core taken from about 580 feet below Piscataway, NJ. The drill runs 24 hours a day. (Courtesy of Janet Kroboth-Weber.)

Figure 19. Dr. Dennis Kent (right) and William Witte, a graduate student, pack sections of core samples for future research at Columbia University's Lamont–Doherty Geological Observatory.

which they show to have been operating for many millions of years, expectations are that the new drilling program with its exceptionally long time frame will uncover longer cycles spanning 2–4 million years, which may be controlled by the earth's interactions with other planets.

Paul Olsen, one of the project's two chief scientists, remarked that "a feature that is evident in these core sections is the strength of these cycles." Earlier studies of ice and sea-sediment cores show that, in more recent geologic times, the 100,000-year cycle appears stronger than would be expected to result from orbital changes, which was thought to be related to the ice. "But," he continued, "we have seen that the signal is very strong during times when there is no ice on earth." The underlying thrust of these studies proceeds from the proposition that to understand how or if mankind is or may be altering current climate, it is essential to know what the climates of the past were and where they can go in the future (13). Additional forcing seems to come from carbon dioxide. If, however, pre-industrial and ice-age levels of CO_2 were in the 200-ppm range, what can the current levels of 350 ppm portend? Should the world be on the way to another ice age as past history appears to call for, or can increasing atmospheric levels of CO_2 force a continuance of the current interglacial or, worse, increase the global temperature? Does the atmosphere hold the answers?

It is worth bearing in mind that whatever set of natural forces were involved, some natural mechanism induced changes in atmospheric CO_2 levels long before mankind embarked on the combustion of fossil fuels in earnest.

REFERENCES

1. Ludlum, D. McW. *Early American Winters: 1604–1820*. The History of American Weather Series. American Meteorological Society, Boston, 1966.

2. Milankovitch, M. Matematische Klimalehre and Astronomisch Theorie der Klimaschwankungen. In: W. Koppen and R. Geiger (Eds.). *Handbuch der Klimatologie, I*. Gebruder Borntraeger, Berlin, 1930.

3. Milankovitch, M. Astronomische Mittel Zur Erforschung der Erdgeschichtlichen Klimate: *Handb. Geophys.* 9:593–698, 1938.

4. Frakes, L. A. *Climate Throughout Geologic Time*. Elsevier Scientific Publishing Company, Amsterdam, 1979.

5. Urey, H. C. Oxygen Isotopes in Nature and in the Laboratory. *Science* 108:489–496, 1948.

6. Emiliani, C. Pleistocene Temperatures. *J. Geol.* 63:538–578, 1955.

7. Hays, J. D., Imbrie, J., and Schackleton, N. J. Variations in the Earth's Orbit: Pacemaker of the Ice Ages. *Science* 194:1121–1132, 1976.

8. Barnola, J. M., Raynaud, D., Korotkevich, Y. S., and Lovius, C. Vostok Ice Core Provides 160,000 year Record of Atmospheric CO_2. *Nature* 329:408–414, 1987.

9. Oeschger, H., Beer, J., Siegenthaler, U., Stauffer, B., Dansgaard, W., and Langway, C. C. Jr. Late Glacial Climate History from Ice Cores. In: J. E. Hansen and T. Takahashi (Eds.). *Climate Processes Sensitivity to Solar Irradiance and CO_2*. Ewing, Series 4. American Geophysical Union, Washington, DC, 1985.

10. Shackelton, N. J., and Pisias, N. G. Atmospheric Carbon Dioxide, Orbital Forcing, and Climate. *Am. Geophys. Union Geophys. Monogr.* 32:303–317, 1985.

11. Genthon, C., Barnola, J. M., Raynaud, D., Lorius, C., Jouzel, J., Barkov, N. I., Korotkevich, Y. S., and Kotlyakov, V. M. Vostok Ice Core: Climatic Response to CO_2 and Orbital Forcing Changes Over the Last Climatic Cycle. *Nature*. 329:414–418, 1987.

12. Lorius, C., Barkov, N. I., Jouzel, J., Korotkevich, Y. S., Kotlyakov, V. M., and Raynaud, D. Antarctic Ice Core: CO_2 and Climate Change Over the Last Climatic Cycle. *EOS Trans. Am. Geophys. Union* 69:681–684, 1988.

13. Olsen, P. Office of Public Information. Columbia University, October 12, 1990.

CHAPTER 2

A BENEFICENT GREENHOUSE

Forty-three to forty-five Fahrenheit degrees is cold: the cold of a refreshing beer on a warm summer day. By comparison, the average global temperature of the air is a pleasantly mild 59°F (15°C). Mild or not, most of us would not swim in water at that temperature—not even on a steamy July or August afternoon. It would be in and out for a quick dip, if at all. At 68–69°F, however, most everyone, swimmers and dunkers, would be in. The 9–10°F increase would make an inviting difference. On the other hand, for those dressed accordingly, a 59–60°F day would be invigorating, a delightful spring day. But what if . . . ? What if global mean surface temperature rose from 59°F to 68°F? Would that be desirable? Sounds fine at first, but the operational term is *average* or *mean*, which hides, as do all measures of central tendency, a good deal of information. To obtain this average, the world's temperature must be taken, which is no mean feat. What will be used for a thermometer and what time of day will the daily temperature be taken; and where will the thermometer be placed—open field, shade of trees, or side of buildings? No single value obtained at any one location will do. Temperatures must be taken in many locations in every country on all continents over 12 months and must include the oceans, which occupy 72% of the earth's surface. What if we miss a day or two, especially in the dead of winter; will that matter? When the temperatures are finally obtained and all the values averaged, a mean surface temperature of 59°F (15°C) is obtained. This is the year-round temperature close to the ground. Would an increase of, say, as much as or as little as 5 or 6°C (9–10°F) be in our best interests? It might. Then again . . . .

Of all the planets in our solar system, the Earth is remarkably unique in possessing so comfortable and hospitable a temperature—a level of comfort unmatched anywhere in our solar system. And it has had this reliably stable climate for the past 12,000–14,000 years, since the last ice age. The temperature and its accompanying climate have been eminently suitable for the growth of all living things. Not so our nearest celestial neighbors, Mars and Venus. Venus is far too hot to support life, and Mars is too cold.

From the recent Viking space probes, we know that the exceptionally thin gaseous blanket girdling Mars is 95% carbon dioxide with little, if any, moisture. In fact, it never rains on Mars. It's so cold there that what little moisture exists is frozen solid. If the Viking probes are correct, anyone contemplating setting foot there should be dressed to accommodate its $-50°C$ ($-58°F$) temperature. High levels of CO_2 are also present in the gaseous envelope encircling Venus. On Venus, however, the 95% concentration of CO_2 occurs in an exceptionally dense atmosphere. And, along with its heavy cover of clouds, there's little escape of heat, which sends its daily surface temperature soaring to a broiling 470°C (878°F). The Moon, our nearest neighbor, has no atmosphere at all. With one-sixth the Earth's gravitational pull, any gases naturally present have long since escaped back into space. Still, sunlight does fall on the Moon. But, without an atmosphere, a portion is reflected and a portion absorbed. Without a heat-trapping mechanism, the absorbed portion quickly escapes and the daily temperature plummets to a chilly $-160°C$. Without an atmosphere, the Moon is a barren, dead place. As for Mercury and Jupiter, they are far too hot, while Saturn, Uranus, Neptune, and Pluto are literally in deep freeze. Thus, in our solar system, the Earth alone is comfortably liveable and in large measure owes its uniqueness to a beneficent atmosphere. Given its remarkable life supporting character, not only is our atmosphere to be respected but the question at issue is whether it can be tampered with. This is quite different from should it be tampered with. Are there lessons to be learned from our increasing knowledge of the Moon, Mars, and Venus? Is there margin for error? With a solar system so vast, so immense, is it possible, even imaginable that anthropogenic activity could adversely affect it? Is our thinking merely self-indulgent egocentricity on our part, or has humankind emerged as a force on a par with natural forces? Can we move mountains and alter our atmosphere with the climate changes these could incur? Perhaps. We shall explore these tantalizations. But we must first consider the atmosphere—that thin girdle of gases that encircles that earth. The earth and the atmosphere, however, cannot be separated from incoming solar radiation; they are inextricably linked.

The sun is hot; approximately 6000°C (10,000°F) at its surface, which is constantly emitting heat energy at the rate of 56×10^{26} calories, or in radiation units, langleys per minute (ly/min). That's 56 followed by 26 zeros: 5600000000000000000000000000! But the spinning, wobbling earth intercepts a mere fraction of this energy and, within the atmosphere, further reductions occur. If it didn't, the oceans would have been brought to a racing boil within 2 seconds and would have boiled away long before Homo erectus made it to an upright position.

These intercepted rays are difficult to visualize, primarily because current theory defines electromagnetism in terms of alternating waves of electric and magnetic fields. Figure 1 portrays the concept of a wave and the idea of a wavelength, which is defined in terms of its length, speed, and

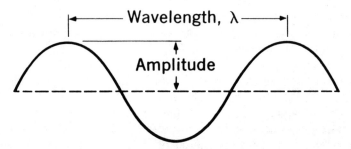

Figure 1. Schematic diagram of a wavelength, λ, from crest to crest. Red or violet light, that is, visible light completes one wave or vibration in the same distance and time that involve two complete waves of ultraviolet light. Hence ultraviolet (UV) light is of shorter wavelength than violet light.

frequency of oscillation. As shown in the figure, distance is measured from crest to crest; frequency is the number of crests passing a given point in a unit of time—most often seconds. The product of wavelength and frequency yields the speed of the wave. The unit of wavelength used to describe electromagnetic waves is the micron (μm), 1/25,000 of an inch or 1/1000 of a millimeter. As we shall see, close to 100% of the sun's radiation falls between 0.15 and 4.5 μm, which encompasses the short-wave ultraviolet, the visible, and the near infrared. Figure 2 displays the complete electromagnetic spectrum from the shortest, most energetic x-rays and gamma rays to the longest, less energetic radio waves.

The spectrum of wavelengths generated by radiating bodies is exemplified by the curves in Figure 3. Here we see the results of black bodies radiating at two different temperaturs (1). A black body is one that absorbs all the radiant heat that falls on it. Thus, as a perfect absorber, it is also a perfect emitter and emits radiation over a broad spectrum of wavelengths—a function of the temperature of the radiating source. The hotter the source, the shorter the wavelength of the induced radiation. And that is what we see in Figure 3. At 6000 K (10,500°F), the radiation is primarily between 0.15 and 1.5 μm, while at 255 K (-18°C or 0°F), it is primarily between 8 and 50 μm. Six thousand degrees is similar to the surface temperature of the sun. It is extremely hot and, as a consequence, emissions are at the shorter, more highly energetic wavelengths, which include the ultraviolet, visible, and near infrared. At the left in Figure 3, the cooler black body radiates at the longer, less energetic wavelengths, well beyond the visible and near infrared. Accordingly, the wavelengths are beyond 5 μm, primarily between 7 and 60 and on to 100 μm. Natural emissions of thermal infrared can be visualized as similar to that emitted by a black body at a temperature of 255 K—equal to the temperature 3.5 miles (18,000–20,000 feet) above the earth into the troposphere. By comparison, the average global surface temperature is 288 K (59°F or 15°C). A black body radiating at 288 K would not differ significantly. The essential point

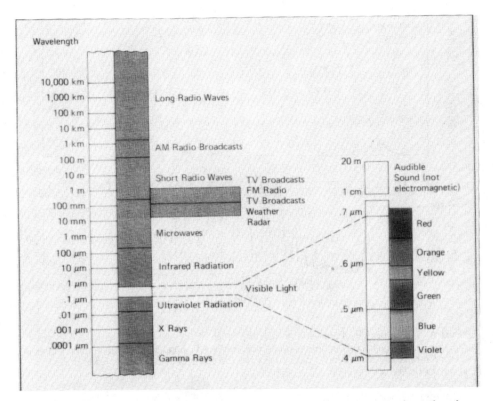

Figure 2. Profile of the electromagnetic spectrum portraying the great range of wavelengths and the fact that the visible band is only a small fraction of the total. (See color section.)

is that the earth is so much cooler than the sun and therefore emits radiation in the infrared region. As we shall see, this type of radiation becomes central to the entire discussion of a potentially warmer world.

Of the total energy emitted by the sun, some 46% is white—visible sunlight composed of seven colors (red, orange, yellow, green, blue, indigo, and violet)—which, in terms of wavelength, extend from red at 0.75 μm to violet at 0.35 μm. Curiously enough, our nervous system, specifically cells on the retina at the rear of the eye, responds to wavelengths between 0.35 and 0.75 μm, and the retina's rodlike cone cells register color. What an odd coincidence. And surely some have wondered why we see a blue, rather than a red or green, sky. As light passes through the atmosphere, a portion is absorbed and a portion scattered or reflected. It is the scattering that accounts for it. And it is the shorter wavelength rays, the blue and the violet, which are scattered widely, giving the sky a bluish hue on clear days. When smog is present with its larger, more absorptive properties, the sky can take on a yellowish hue.

An additional 48–49% of incoming radiation is of the near infrared. We do not see it, but it is nevertheless sensible; it is felt as heat. The remaining

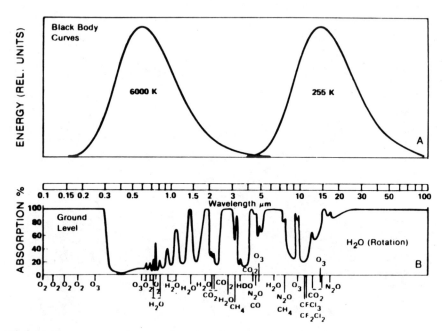

Figure 3. The upper set of curves describes the range of wavelengths encompassed by the radiation of a black body whose surface temperature is 6000 K, as well as one whose temperature is 255 K. The lower curve describes the absorption characteristics of the major atmospheric gases.

5% is similarly invisible but is on the shorter side of violet and is referred to as ultraviolet, beyond 0.35 μm. It is not felt unless and until we've received so large a dose that our skin becomes erythematous (red or sunburned).

Radiation hurtling through space does so in straight lines and must penetrate an envelope of gases before striking the surface of the earth. For 300 miles from ground zero, planet Earth is sheathed in an envelope of gases—a veritable protective cocoon. Figure 4 attempts to show this layering in a cutaway sketch, while Figure 5 is a photograph taken from space. Here the stratosphere is seen in blue and the troposphere in white and pink. Before the energetic particles of electromagnetism strike the earth, they must run the gamut of this multilayered atmosphere.

The atmosphere has conveniently been divided into layers on the basis of temperature gradation. Figure 6 shows the organization of the atmosphere with respect to temperature. When air rises, it expands and cools; thus air temperature reasonably should be expected to decrease with increasing altitude. But temperature recording instruments aboard radiosonde balloon and sounding rockets have found that, although temperatures do drop precipitously beyond the troposphere, the decline is not

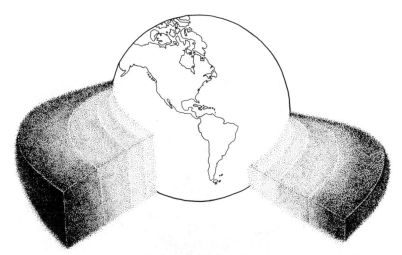

Figure 4. Schematic illustration of the atmosphere girdling the earth, showing its characteristic layering.

Figure 5. This photograph, a cross section through the Earth's atmosphere over the Pacific Ocean, was taken from the Space Shuttle *Challenger*, on October 10, 1984, from an altitude of 123 nautical miles. Stratosphere appears blue, while the troposphere appears white and pink. (Courtesy of M. R. Helfert, NASA–Johnson Space Center, Houston, Texas.) (See color section.)

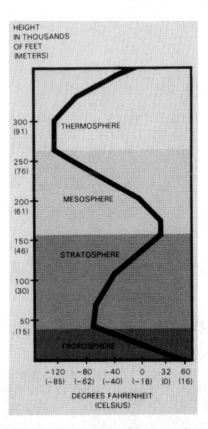

Figure 6. A profile of the changes in temperature of the atmosphere with increasing altitude.

linear. As noted in the figure, the changes, seeming to shift haphazardly, in fact, do so for good reason.

Recall that 0°C is freezing (32°F). Thus through the troposphere, the layer closest to the ground, the temperature falls directly with increasing altitude. As we reach the top of the troposphere, and the bottom of the stratosphere, the temperature has cooled considerably to a very cold −50°C. It then slowly reverses itself and begins to warm. As the 30-mile high level is attained at the top of the stratosphere, it has warmed to a cool 5°C—approximately 41°F, a little cooler than most household refrigerators. Through the mesosphere, it cools again to about −100°C; and into the aptly named thermosphere (ionosphere) another reversal occurs, this time to about 45°F, just about the temperature of most household refrigerators.

Troposphere, from the Greek *tropos*, meaning turning or changing, refers to the widely varying local weather conditions that occur here. The occurrence of our weather in this layer is inextricably bound to the presence of water vapor. Almost all the water in the atmosphere, in its liquid and gaseous phases, is contained in the troposphere. Clouds, another feature of the troposphere, are also tied to the presence of available water, water below its freezing point.

Between 15 and 50 kilometers (9–30 miles), temperatures rise, placing warmer air over cooler. This is a stable arrangement and little vertical mixing occurs in the stratosphere. The warmer air thus acts as a lid on convection in the troposphere. If there were no warmer layer, convective systems would rise higher into the stratosphere and, above, clouds would form at far greater heights and the climate of the earth would be much different. The stratosphere is additionally unique in that, within this band, there is an encircling band of ozone. Above the stratosphere is the mesosphere, in which temperatures drop markedly. At an altitude of 50 miles (80 km), the temperature is in the range of $-100°F$ ($-73°C$). Above the mesosphere, temperatures rise precipitously into the well-named thermosphere, where temperatures reach levels of 2200–2300°F (1200–1200°C). Heating in this layer is the consequence of intense incoming high-energy ultraviolet radiation that smashes into molecular oxygen (O_2) and nitrogen oxide (NO_x), releasing electrons and, in the process, heating the surrounding atmosphere. With (their) electrons removed, these chemical species become positively charged ions—hence, the second designation of ionosphere for this warm layer. This layer then carries two names, each indicating one of its two major characteristics.

Nevertheless, less than 0.0001% of the atmosphere is in the thermosphere. Beyond these layers and above 200 kilometers (180 miles) are the exosphere, magnetosphere, and Van Allen belts of highly charged particles. As indicated, radiation travels in straight lines and, when striking objects, such as molecules and atoms of a variety of chemical species dust, earthly detritus, and clouds, it will pass through or be held back, deflected, and absorbed. Those objects through which it passes are said to be transparent and those that hold light back are opaque. Opaque materials absorb heat and become warm. That's another key. They can also reflect and reemit their heat—still another key.

The atmosphere is transparent to solar radiation: that is, the more energetic short wavelengths readily pass through. A portion strikes the earth, is absorbed by it, and warms it. The earth, being a reflecting body and far cooler than the sun, reradiates energy, but at the longer wavelength infrared (IR), and herein hangs the tale. Actually, the take hangs on the chemical composition of the atmosphere. Table 2.1 displays the mixture of its ingredients. Clearly, well over 99% of the air we breathe consists of nitrogen, oxygen, and argon. Curiously enough, these major gases absorb insignificantly in the infrared. While constituents such as water vapor (H_2O), carbon dioxide (CO_2), ozone (O_3), and methane (CH_4), shown in Figure 7 and present at parts per million concentrations, are transparent to incoming short-wave radiation but opaque to the long-wavelength infrared. With these strong IR absorbers, a little goes a long way; and it is here that a delicate balance could be compromised—or so it has been suggested.

Table 2.1
The Gaseous Composition of the Atmosphere

Constituent Gas	Chemical Formula	Abundance (Volume)[a]	
Nitrogen	N_2	78.08%	
Oxygen	O_2	20.95%	
Argon	Ar	0.93%	
Water vapor	H_2O	Variable (%-ppmv)	
Carbon dioxide	CO_2	340 ppmv	(0034%)
Ozone	O_3	0.03–10 ppmv	(0.00003–0.0010%)
Methane	CH_4	2 ppmv	(0.2%)
Neon	Ne	18 ppmv	
Helium	He	5 ppmv	
Krypton	Kr	1 ppmv	
Xenon	Xe	0.08ppmv	
Hydrogen	H_2	0.5 ppmv	
Nitrous oxide	N_2O	0.3 ppmv	
Carbon monoxide	CO	0.05–0.2 ppmv	
Ammonia	NH_3	4 ppbv	
Nitrogen dioxide	NO_2	1 ppbv	
Sulfur dioxide	SO_2	1 ppbv	
Hydrogen sulfide	H_2S	0.05 ppbv	

[a] ppmv and ppbv are parts per million and parts per billion by volume.

Much of the reradiated energy, reflected and emitted by the earth, is absorbed—trapped in the stratosphere by carbon dioxide, ozone, and methane and in the troposphere by water vapor. Except for strong absorption by ozone at 9.6 μm (note once again Figure 7), the region 8–12 μm allows terrestrial reradiated infrared to escape back into space via this atmospheric "window." But it is relatively narrow and restrictive. As we shall see, this "window" has become even more restrictive as a consequence of anthropogenic activity.

Having absorbed the infrared from the earth, the quartet of trace gases reemits it in all directions, and a portion of this heat once again strikes and is absorbed by the earth. Here then is the touchstone: the presence of these strong IR absorbers, at trace levels, is responsible for maintaining the global mean surface temperature at 15°C—some 33°C above what it would be if they were absent. Consequently, these gases must be seen as a highly beneficial, naturally protective life support system. This constellation of events, shown in Figure 8, is in fact the reason why Earth, of all the planets in our solar system, has a humanly hospitable climate. But the underlying element is the Earth's energy balance.

Between the incoming solar radiation and the radiation reflected from the earth and atmosphere, an energy balance is struck. If absorbed solar

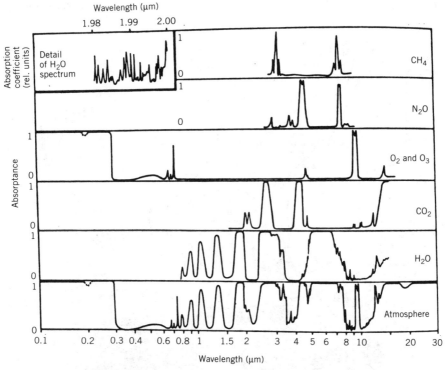

Figure 7. The absorption peaks and specific wavelengths of each of the atmosphere's five major constituent IR-absorbing gases are displayed. Each is separated from the total atmospheric spectral absorbance. Fitting them together would yield the complete spectrum.

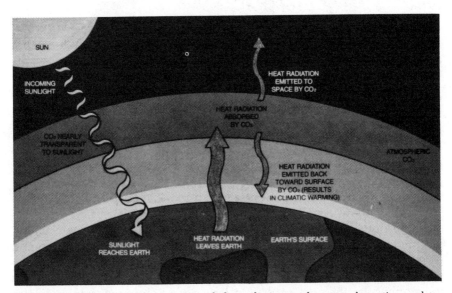

Figure 8. Diagrammatic representation of the radiation exchanges—absorption and reflection—which result in the beneficent "greenhouse" effect. (See color section.)

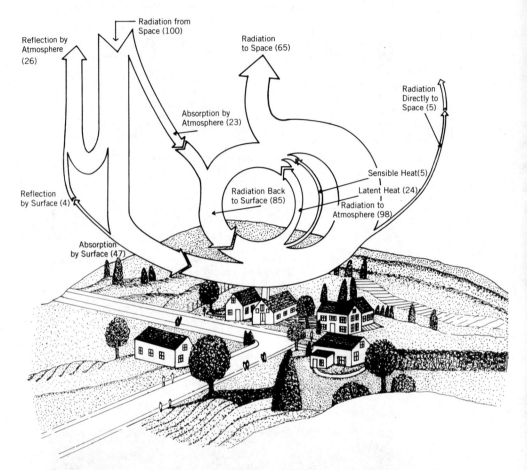

Figure 9. Diagrammatic representation of the globally averaged heat balance of the earth. Incoming and outgoing radiations are shown drawn to scale to indicate the relative contribution of each major cpmponent. (Reproduced with permission from *Carbon Dioxide Review*, copyright © 1982 Cambridge University Press, Cambridge, England.)

radiation is balanced by the range of reradiating sources, the global energy system will maintain the stable mean surface temperature that it has enjoyed over the past 10,000 years. It is the total balance of energy that produces the 15°C global mean temperature.

It works this way. As light waves pass into the atmosphere, a portion is scattered as the waves bounce off molecules of air, water, and dust. Another portion is scattered and re-reflected by clouds, and a substantial portion reaches the earth, striking everything in its path—trees, homes, people, animals, cars, lakes, and hills—to be absorbed and reflected, depending on the type of surface. Much of the short-wave length radiation is transformed by the cool earth into longer wave infrared, which excites molecules in its path to increased activity—warmth. Figure 9 illustrates the energy pathways. On the basis of 100 units of solar radiation entering at the top of the atmosphere, it has been estimated, as the diagram shows, that 30% of the short-wave radiation is reflected back to space by clouds, particulate matter, and gases. The surface and atmosphere absorb the rest. The essential facet of the balancing is that the earth gives up heat and cools to a humanly comfortable temperature. Therefore if incoming solar radiation—insolation—accumulates in the atmosphere beyond the point of balance, for whatever reason, global temperature should rise, unevenly to be sure, but it should rise, if only because radiation cannot dissipate. The windows are blocked; it has no place to go.

At the heart of the potential warming issue are small temperature increases—1, 2, 3, perhaps 4°C. These numbers are seemingly small, when we consider the 14–15°C swing normally experienced between winter and summer in the northern hemisphere. But these small increments are averages, which mean that, at certain locations around the world, increases could be as much as 10 or more degrees, while at other locations, temperature increases could be less than 1°C. The crux of the issue is this: Is the energy balance shifting? Has the heavy reliance on burning fossil fuel overburdened the atmosphere with IR-absorbing gases to the point that heat is accumulating, producing what climatologists refer to as radiative forcing? If so, over time, the pleasantly mild 15°C mean surface temperature could rise, adversely affecting the world as we know it.

The sequence of events in which short-wave radiation is absorbed by the earth and reradiated as longer wavelength IR that is trapped in the atmosphere, which is depicted in Figure 8, was referred to as a greenhouse effect in 1827 by the French mathematician Jean Baptiste Fourier. And, in 1861, John Tyndall, a British physicist experimenting with radiant heat and gases, noted the differences in absorbance of atmospheric gases and likened them to a greenhouse. Why greenhouse, and is this a paradigm for the atmosphere?

As Figure 10 shows, the glass of a greenhouse permits the passage of visible short-wave radiation but absorbs the longer wavelengths. The light falls on the objects inside, warming them and raising their temperature. The plants, pots, tools, and flowers then reradiate part of that energy, and the temperature rises. The high temperatures developed in a greenhouse, however, are not attributable to the absorption of infrared radiation by glass. Fleagle and Businger (2) inform us that, in 1909, R. W. Wood experimented with greenhouses. Using two model greenhouses, one covered with glass and one with rock salt (sodium chloride)—which is transparent to both short- and long-wave radiation, so that trapping of radiation and consequent heating does not occur—he found that both greenhouses attained similar high temperatures. The temperature in a greenhouse increases above that of the outside air, because its iron-containing glass roof

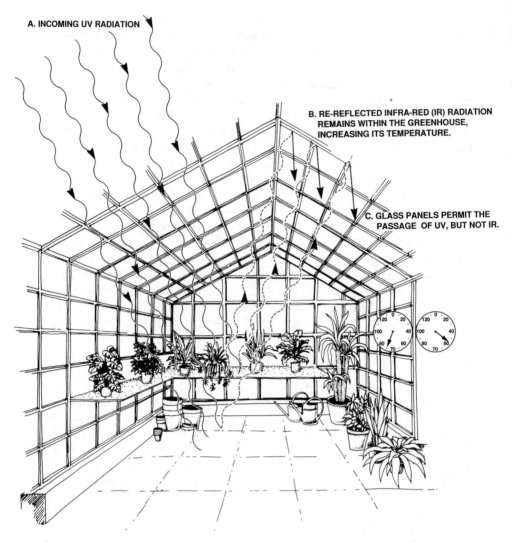

A. INCOMING UV RADIATION

B. RE-REFLECTED INFRA-RED (IR) RADIATION REMAINS WITHIN THE GREENHOUSE, INCREASING ITS TEMPERATURE.

C. GLASS PANELS PERMIT THE PASSAGE OF UV, BUT NOT IR.

Figure 10. Key to comprehending the underlying principle of a greenhouse or "hot house."

and walls prevent convection currents from dissipating the heat outward. This purely physical effect is far more important to the working of a greenhouse than the absorption of long-wave radiation. According to Wood, the effect is four to five times as great as the absorbtion of long-wave radiation by the glass. In addition, greenhouses can be vented to prevent excessive warming, which would cause the flowers and plants to wilt and/or die—a common occurrence in early greenhouses. Consequently, the analogy between a greenhouse and the atmosphere can only be pushed so far. Ann Henderson-Sellers and P. J. Robinson have suggested that a better analogy would be that of a leaky bucket or buckets. Taps in the buckets permit a constant flow of heat onto the buckets, but one bucket with less infrared absorbers acts as a bucket with a large hole, permitting rapid emptying of heat. A second bucket with a large concentration of infrared absorbers acts as a bucket with a small hole from which emptying occurs slowly and heat builds up. It is, however, particularly important to realize that the natural "greenhouse" effect is a highly salutary condition found only on planet Earth. Rather than a greenhouse, Venus is a hothouse—a very hot house—and Mars is a veritable ice house. Why the pressing concern for a problem that has mistakenly come to be known as the greenhouse effect? Concern, as noted earlier, focuses on the very real possibility of upsetting what may be a delicate gaseous balance. If underlying theory is correct, these emissions would result in excessive heating within the greenhouse, enough to wilt the flowers and other growing things. Thus concern should not be for the greenhouse effect—which, in fact, is entirely beneficent—but rather for excessive warming within the greenhouse—our atmosphere—as a consequence of human activity.

Is there anything of value to be gleaned from our increasing knowledge of the atmospheres of Mars and Venus, especially Venus, which is so similar to Earth? Venus is a bit closer to the sun and would be expected to be warmer. Could that extra bit of warmth have initiated a chain reaction in which CO_2 was released in abundance from carbonate-containing rock, and, as additional CO_2 was released, the planet grew increasingly warm, producing the current runaway greenhouse? It's worth considering, and the key idea is warmth. The fact is, with our ability to monitor and measure levels of each of the four major trace greenhouse gases, a number of our best scientists have become fidgety with what they see as significantly increasing concentrations. The idea that small changes can exert large effects is not lost on them—given the enhancing capabilities of positive feedback effects, which could increase warming effects considerably.

From ancient Antarctic ice-core data, we now know that the trace gases occur naturally and that they also fluctuate naturally, but the long, historical records revealed by the cores also indicate that the level of CO_2, for example, never exceeded 290 parts per million—0.029%—until the industrial age. Should the balance shift, a new balance would take effect and would be accompanied by a higher global surface temperature—perhaps 17, 18, or 20°C (62.6, 64.4, or 68°F). Again, would this world be more hospitable, more comfortable? Global warming could be viewed as a (natural?) process in which an energy balance is maintained or conserved. In this instance, the handiwork of humans may be sufficiently powerful to force a shift. The problem for humankind is that a new, warmer balance, warmer earth may prove unfriendly for growing things in different areas of the planet—especially for those growing things that have evolved in temperate climates with all the benefits that entail. In the long run, human beings may prove inimical to their best interests. But let us look closer at the trace gases that loom so large in our world. Perhaps the data do not support so pessimistic a view. Then again . . . .

CARBON DIOXIDE

Not only is carbon a constituent of all living things, but life as we know it could not exist without it. The major source of carbon for all living things is the atmosphere, and the principal way in which carbon moves from that inanimate sphere to the living biosphere is via photosynthesis. As generally conceived, photosynthesis is a multistep process in which sunlight reacts with chlorophyll, the green pigment of plants, and together, they react with carbon dioxide and water as follows:

$$\text{sunlight} + 6CO_2 + 6H_2O \rightarrow C_6H_{12}O_6 + 6O_2.$$

This formulation shows the outcome of the reaction as the formation of carbohydrate (glucose, a sugar) and the production of oxygen. In Figure 11,

Figure 11. Oxygen-rich bubbles on the leaves of a submerged green plant show that photosynthesis is taking place. (Courtesy of Quesada/Burke, New York.) (See color section.)

oxygen-rich bubbles are shown being emitted from the surface of the leaves of a green marine plant—a unique graphic representation of the end product of photosynthesis.

When such fossil fuels as coal, oil, and peat (all carbonaceous materials) are burned, the major product of this combustion—respiration, the reverse of photosynthesis—is carbon dioxide. Respiration proceeds this way:

$$C_6H_{12}O_6 + 6O_2 + \text{heat} \rightarrow 6CO_2 + 6H_2O.$$

It is this release of CO_2 into the atmosphere that is at the core of the global warming controversy. Over the past 150 years, inordinate amounts of carbon dioxide from burning oil and coal to fuel the engines of society have been poured into the atmosphere, thereby returning to it the atmospheric carbon that had been removed hundreds of millions of years ago by trees and plants in their processes of photosynthesis. Thus in a sudden, relatively brief span of time, less than three generations, the contemporary atmosphere has become taxed with vast additional quantities of a strong IR absorber. In the process of photosynthesis, green plants absorb some 230 billion tons of CO_2 yearly. Similar tonnages are released via respiration. This process of absorption and emission has been a steady-state process for at least the past 14,000 years. Suddenly, over the past 150 years, the combustion of coal and oil in ever increasing quantities has weighted the respiratory side of the equation to a degree some believe is producing an imbalance. (Given that cleaner burning of fuels is the equivalent of adding additional CO_2 to the atmosphere, curbing pollution could be counterproductive and may require alternative solutions.) In 1988 alone, an additional 5 billion tons of carbon were released into the atmosphere. If these numbers seem large, consider that, according to Table 2.1, CO_2 is present in the atmosphere at a level of 0.03%—a trace gas; compared to oxygen and nitrogen, it is minuscule. Nevertheless, 0.03% represents some 750 billion tons of atmospheric CO_2. It is into this level of natural CO_2 that artificial or unnatural additions are being poured. Remember, an increase in the amount of atmospheric CO_2 tends to close the infrared "window," because there are more molecules of CO_2 to absorb the radiation, thus blocking its passage back to space. Thus it is entirely reasonable to question whether such additions of this gas alone can be managed by the natural carbon cycle, and for how long, without adversely shifting the natural heat balance. We fail to answer these questions at our peril.

Fortunately, this is not a new question. It was considered as long ago as 1896, when the outstanding Swedish chemist Svante Arrhenius not only posed the question but calculated the temperature increase to be expected for a doubling of CO_2 in the atmosphere (3). With advancing and increasing industrialization, the idea of a doubling CO_2 content has become more than an academic exercise and has occupied the considerable time of a

number of eminent scientists. Thus as part of the massive scientific effort
of the last International Geophysical Year (IGY), 1957–1958, projects for /
the measurement of atmospheric CO_2 were undertaken.

[Charles David Keeling and his colleague Robert B. Bacastow, of the
Scripps Institution of Oceanography, in conjunction with the National
Oceanic and Atmospheric Administration (NOAA), established CO_2 monitoring stations on the summit of Mauna Loa on the island of Hawaii and at
the South Pole Station of the U.S. Antarctic Program. These sites were
chosen because of their isolation from industrial areas as well as local air
pollution, but they permit collection and analysis of well mixed air—an
important consideration for appropriate air sampling, if it is to represent
all air at the concentration in the middle of the troposphere.

From 1958 onward, Keeling has collected air samples at the Mauna Loa
Observatory and has analyed them for their CO_2 levels. Figure 12 is the
most recent and up-to-date rendering and is among the most quoted and
reproduced pieces of scientifically and politically scrutinized data of the
past decade. One thing is clear: the trend is upward, continuously upward.
What we have here are collections of mean monthly CO_2 values. Actually,
these monthly averages are the end-product of hourly, daily, and weekly
averages. Thus each point on the graph is the residual of literally thousands of analyses. Furthermore, after 1974, NOAA and Scripps had dupli-
cate infrared gas analyzers running in the main building. And even though
the air inlet was the same for both analyzers, there were differences in
readings between instruments. But both instruments were obtaining CO_2
data, which required that both groups work closely together to be certain
the data reported were accurate. The seasonal fluctuations, which are
superimposed on the rising CO_2 trend line, are due to the uptake of CO_2 by
metabolizing plants and trees in spring and summer in the northern hemisphere—photosynthesis—a net storage of carbon and its withdrawal from
the atmosphere, and an increase in atmospheric CO_2, with its release via
respiratory activity in the fall and winter. The air samples are sufficiently
sensitive to reflect these fluctuations.

The fluctuations are dominated by northern hemisphere vegetation because of the far greater land mass in the northern hemisphere and because
there is less seasonal change in plant activity at the equator. Thus in Figure
12, we see that each open square represents 1 month's average CO_2 level,
and each leg represents a half-year's readings. The descending leg represents photosynthetic activity and the ascending leg the dominance of
respiratory activity. These seasonal variations appear at all clean air
stations. At the South Pole station, the upward CO_2 trend closely parallels
that at Mauna Loa, but the seasonal fluctuations are less steep as a consequence of its isolation from plant life. The central trend line represents the
yearly averages with seasonal variation removed. Although these data
points are from a single station, they are typical of the steady rise in
atmospheric CO_2 obtained over the past quarter of a century at the other

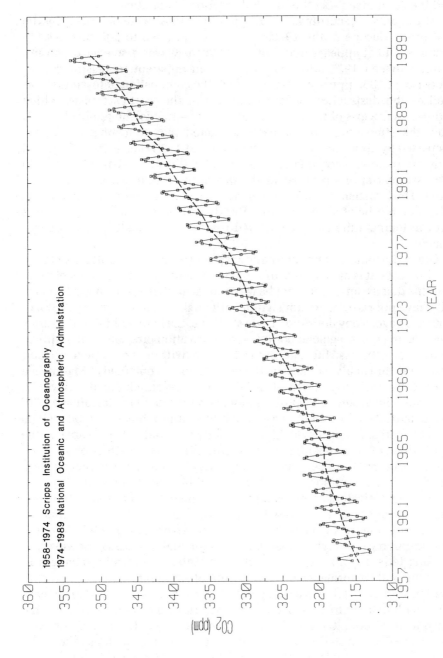

Figure 12. Atmospheric CO_2 concentrations from the Mauna Loa Observatory, Hawaii, 1958–1989. (Courtesy of NOAA/Climate Monitoring and Diagnostics Laboratory. Boulder, CO.)

NOAA stations: Point Barrow, Alaska; Cape Matatula, American Samoa; and the Amundsen–Scott South Pole Station, Antarctica.

At the outset, recordings of 315 ppm (0.0315%) were obtained. By the end of the decade of the 1970s, the level had risen to 335 ppm—a 6% increase. But it appears that around 1973 there was a steeper climb upward. Between 1975 and 1989, there is an apparent increasing rate of increase—to 350 ppm—primarily due to the combustion of fossil fuels, as well as the destruction of large portions of the world's forests, which represents the loss of photosynthetic activity—that is, the removal of CO_2 from the atmosphere. It has been estimated that atmospheric CO_2 has increased by approximately 20% over the past 150 years—since the rise of industrialization—and that more than half of this increase has occurred in the past quarter of a century, as the rate of fossil fuel consumption escalated. Thus Arrhenius' original concern about a doubling of atmospheric CO_2 may not be at all unrealistic. But does it portend an upward global shift in temperature as he calculated it would? We shall look at the evidence.

Given the national and international policy implications attached to the Keeling data, it is appropriate to consider its collection and interpretation. But this discussion should not be taken to mean that atmospheric levels of CO_2 have not risen or are not significantly higher than they were 15, 30, or 100 years ago. They are. And, as we shall see, they are far higher than they were during the last glacial period—an observation pregnant with implications. Nevertheless, the data must be seen within the context of their collection and analysis. Clearly, the measurements obtained at Mauna Loa are the largest daily series of airborne CO_2 levels obtained anywhere, anytime. The Mauna Loa Observatory is on the island of Hawaii at 19.50° north latitude and 155.6° west longitude, approximately 3400 meters (11,000 feet) above sea level. Figure 13 shows the rather barren area on the north slope of Hawaii's highest mountain. Since its construction in 1956, the observatory has been operated by NOAA. The site was picked for a number of considerations, as noted earlier; thus the data obtained should be reliable indicators of atmospheric levels unencumbered by extraneous CO_2 contributions. The 88-foot tower, with its set of four air intakes, is shown is Figure 14. These towers are approximately 570 feet from the CO_2 infrared gas analyzers maintained in the main building. However, as Keeling tells us, "in spite of efforts to maintain uniformity, changes in instrumental performance and procedure of sampling, analysis, and calibration were inevitable in a program lasting over 23 years" (4). This remark dealt with data up to 1980. By 1990, it had run continuously for 33 years. Despite all efforts, it was difficult to separate out contributions from volcanic activity, vegetation, and artificial contributions such as auto emissions. It is also to their credit that control gases of known CO_2 levels were used regularly and frequently to assure that the readings obtained with atmospheric samples were valid and accurate. At times, the rubber dia-

Figure 13. Site of the Scripps/NOAA Laboratory 3400 meters above sea level on barren Mauna Loa. In addition to measuring CO_2 levels, measurements of aerosols, ozone, and solar coronas are also made here. (Courtesy of C. D. Keeling, Scripps Institution of Oceanography, La Jolla, CA.) (See color section.)

Figure 14. The Scripps/NOAA site on Mauna Loa, showing (at left) the 88-foot high tower with it CO_2 collection intakes at the top. (Courtesy of C. D. Keeling, Scripps Institution of Oceanography, La Jolla, CA.)

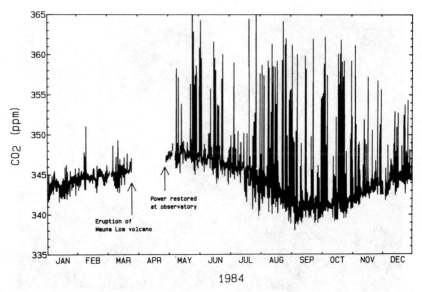

Figure 15. Hourly average CO_2 concentrations for 1984. Power to the observatory was cut by a lava flow in late March. Increased variability of CO_2 is evident after power was restored 1 month later. (Courtesy of K. W. Thoning, Cooperative Institute for Research in Environmental Sciences, University of Colorado, Boulder.)

phragm pumps used to suck in air from the intakes were responsible for producing persistently high CO_2 levels. It is also well to recall that CO_2 was not read directly but had to be derived from voltage recordings. Here too was another potential source of error. For several years, high readings were obtained at noontime. Eventually, it was discovered to be the result of an unauthorized vehicle approaching the area. When a locked gate was erected, the high reading vanished. Thus the data must be scrutinized continuously to be certain that they represent only natural air uncontaminated by local contributions. The task is daunting. Earlier, it was noted that the points in Figure 12 represented monthly averages of recorded CO_2 levels. They do; but the monthly averages are themselves averages of hourly and daily recordings. Figure 15 shows actual hour-by-hour recordings, including a break in the data due to a loss of power. Interestingly, there are two gas analyzers side by side running continuously. One is a NOAA instrument and one is a Scripps instrument. Both obtain their air samples from the same inlet port but do not obtain the same results. Consequently, teams from NOAA and Scripps review the data constantly; in spite of the difficulties as the decades pass, collection and analysis manifestly improve. With data so fundamental to estimates of climate change, it is essential that it be beyond reproach. They are.

THE RECORDS IN THE ICE

The summer of 1988 was scorching in the United States, and not just for a day or two. The smothering weather continued unabated; Iowa, Kansas, and Nebraska, the bread-basket states, were sweltering through the worst drought in 50 years, the possibility that a global warming was at hand loomed large. And in Washington, DC, James Hansen, an eminent climatologist and director of NASA's Goddard Institute for Space Studies testifying before a Senate committee, responded to a question saying: "It's just a logical conclusion that the greenhouse is here." He may have been correct; then again, perhaps not, but the unremitting heat lent credence. He really didn't mean that the greenhouse was here; it's always been. He really meant that there was trouble in the greenhouse: there was excessive heating within the greenhouse and it was affecting our climate—and he had data to prove it. His data, displayed in Figure 16, depicts the annual mean

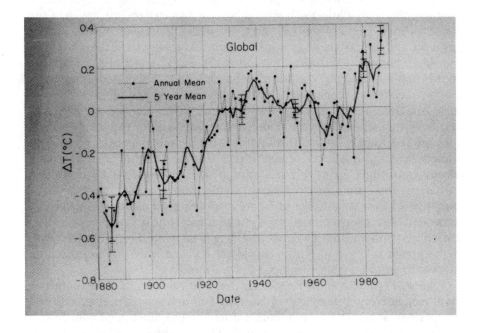

Figure 16. Observed global temperature change between 1860 and 1985, based on data obtained at meteorological stations, but uncorrected for urban heat island effects. (Courtesy of James E. Hansen, Goddard Institute for Space Studies, New York.)

change in temperature, along with the 5-year mean between 1880 and 1985. Admittedly, there is a rising curve from 1880 to 1940, a decline to 1965, and a steady rise thereafter. As Hansen told the senators on the Energy Committee, for the first half of 1988 global temperatures had risen 0.4°C relative to mean temperatures for the period 1950–1980, and he was 99% certain that the accumulation of greenhouse gases (the use of the plural gases is crucial and should be remembered) was responsible for the warming trend. That degree of certainty turned out to be another "shot heard around the world." But, as he explained, the temperature from January to May of 1988 had exceeded three times the standard deviation of that 30-year period, which in statistical terms means there is less than a 1% chance that natural effects were responsible for the change. Statistical confidence was behind his 99% certainty that something other than nature was inducing a warming. Something, but what?

He, along with climatologists of the Climatic Research Unit, University of East Anglia, The Meteorological Office, London, and NOAA believed that "understanding the climate's response to changes in forcing is essential if we are ever to forecast future climate change." They calculated that, by 1987, the global mean temperature was 0.33°C above the 1950–1979 average, and that, while 1987 was the warmest year of record, seven of the eight warmest years of the century occurred in the 1980s (5),(6). Such a level of increase, as small as it seems to be, is greater than the normal or expected: hence the question that follows is: Is this a signal of a warming trend? Jones and Wigley are mindful that the reliability of their time series can be questioned "because the spatial coverage, even at best, is less than 75%, and because the coverages change with time." They go on to remark, however, that when viewed in light of recent ideas of the causes of climate, they are "extremely interesting," in that "the overall change is on the right direction and of the correct magnitude." But they demur; "the relatively steady conditions," they tell us, "maintained between the late 1930s and 1970s requires either the existence of some compensating forcing factor or possibly a lower sensitivity to greenhouse gas changes than is generally accepted." For future temperatures to be forecast, it is therefore essential that past temperatures be known. Proxy data were found in ice cores drilled from Arctic and Antarctic ice sheets. The ice cores suggested an answer to what may have induced the warming.

Figures 17 and 18 show a record of 160,000 years for CO_2 and temperature, respectively. It is evident that, long before humankind began to roll up its sleeves, as it were, both temperature and CO_2 fluctuated naturally. It is also apparent that temperature and CO_2 were high 140,000 years ago, steadily declined through the period we associate with the long ice age, and then rose again some 15,000 years ago. Evident too is the indication that the rises have become increasingly steep over the past 150 years. These are important pieces of information. Especially pertinent is the evidence that CO_2 levels appeared to fluctuate as much as 100 ppm natu-

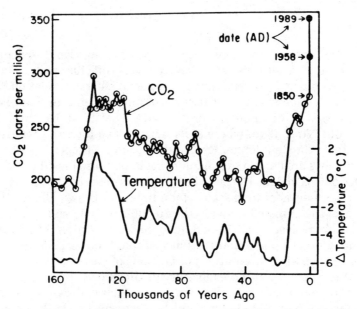

Figure 17. Vostok ice-core data suggest a link between CO_2 levels and mean surface temperature. Indications are that orbital changes are involved, but whether CO_2 leads temperature or temperature leads CO_2 remains to be determined. An important question is: Does it matter? Of additional concern is the natural variation of CO_2 from 190 ppm during cooler periods to 280 ppm during warmer periods. This may be a key.

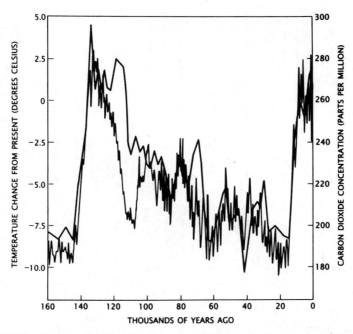

Figure 18. The intimate relationship between actual CO_2 levels (ppm) and temperature change (degrees Celsius) from the present is evident. But how is it to be interpreted?

61

rally. Furthermore, if these two trend lines are superimposed on one another, a new interpretation literally leaps off the page. Perhaps, however, we ought to reverse the positions. What interpretation do we choose? Did CO_2 force temperature shifts or did temperature changes force CO_2 shifts or . . . did neither force either? That is, did a third factor provoke both? And, if so, where is its record to be found? There is little question that climate (temperature) and CO_2 are in it together—that they are inextricably linked. But to conclude that one caused the other was to fly in the face of lack of data. Hansen himself speaks of the confounding effects of century and decadal time scales. John Imbrie of Brown University and many others see climate forcing in the 20,000-, 40,000-, and 100,000-year Milankovitch cycles, while Robert A. Berner and Antonio C. Lasaga of Yale, among others, champion the Phanerozoic time scales in which climate forcing occurs over hundreds of millions of years.

From the scientists at the Laboratory of Glaciology and Geophysics of the Environment, near Grenoble, France, and their colleagues at the Institute of Geography of the Academy of Sciences of the USSR, Leningrad, who together obtained the ice cores with their 160,000-year climate record, we learn that, for the most part, the earth have been a cold place (7–9).

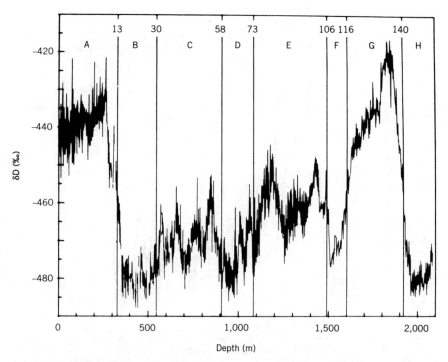

Figure 19. Vostok ice cores also revealed a continuous deuterium profile over the past 160,000 years, which appears to parallel temperature changes over the same period.

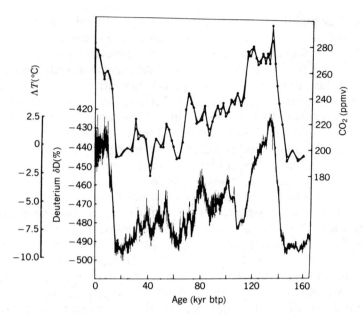

Figure 20. The deuterium levels over time parallel the CO_2 record and suggest that temperature and CO_2 are intensely interactive. The role of orbital forcing is also documented.

There have been ice conditions for over 100,000 years and warmer periods for less than 10,000. And it is over the past 10,000 years that human development has flourished and flowered. Figures 19 and 20 show the continuous deuterium record (content) of the Vostok ice cores, along with the CO_2 variations over 160,000 years. The temperature curve encompasses a shift of as much as 9°C over that period, but an average is far less. It is tempting to superimpose one curve on the other and to assume that temperature variations followed changing CO_2 levels, as a cause–effect relationship. But such temptations must be resisted. For as Barnola of the Grenoble group suggests, "the CO_2 changes . . . are well correlated with the Antarctic temperature record . . . on the same core. Such a high correlation would be expected if CO_2 plays an important role in forcing the climate. The results also suggest a different behavior of the relative timing between CO_2 and Antarctic climate changes, depending on whether we proceed from a glacial to an interglacial period or vice versa. This may have implications concerning the relation between cause and effect . . . but more detailed measurements . . . are required before firm conclusions can be drawn. This does not mean that a cause–effect relationship does not exist. It does mean that it is premature to suggest that it does" (10). Nevertheless, there is an extremely high correlation between the two events, which suggests that CO_2 is surely involved in climate forcing. Most recently, Tom Wigley and Sarah Raper, of the University of East Anglia's

Climatic Research Unit, analyzed the record of global temperature from 1872 to 1982 and, with it, analyzed a range of time scales from decadal to a thousand years and compared them with the pattern of temperature variations observed since the 1860s, when instrumental records began. They found that the pattern of these variations was similar to the actual pattern of changes observed in recent decades. But they also found, and perhaps of greater significance, that the variability of the model on a time scale of centuries is much less than the observed trend of 0.5°C over the past 100 years. Current models predict that, for a doubling of the amount of CO_2 in the air, global warming should be between 1.5 and 4.5°C. If, however, the warming over the past 100 years is seen not as a return to a former warm period but as a greenhouse "signal" due to a 25% higher atmospheric level of CO_2, over that same period, then the implication is a rise in temperature of from 1.3 to 2.0°C for a doubling. However, when Wigley and Raper allowed for the climate's natural variability, the observed warming corresponded to a range of 1.0–2.9°C (11). This appears to strengthen the consistency between model predictions and observations. Apparently, the earth has warmed faster over the past 100 years than can be explained by any natural climatic fluctuations. Nevertheless, they maintain that, while a clear unambiguous signal of a greenhouse warming remains elusive, the observed trend is highly significant statistically. Obviously, we have cycles within cycles to contend with and these require sorting out. Perhaps they're all involved.

The above noted uncertainty notwithstanding, there is yet another piece of data that must be added to the still incomplete climate jigsaw; it is inconceivable that large quantities of CO_2 will not be added to the atmosphere for years to come. While its limits vary on both the available reserves of fossil fuels and its rate of usage, there are upper and lower bounds within which to speculate—otherwise known as ballpark estimates. Thus if half of the known reserves are burned over the next 100 years, the atmospheric CO_2 level could rise to 1.0%: that is, 1000 ppm, three times the current level. If usage only doubles, as past calculations have predicted or used as a basis for discussion, then the current 350 ppm could rise to 600 ppm or there about. For this to be permitted to occur would be an unconscionable form of Russian roulette. Yet it is difficult to imagine arriving anywhere near these levels in the presence of ambiguous signals. Nevertheless, the question that follows is: How long can natural systems sustain such excesses, and what is our margin for error? Is there a margin? Again, without answers, our future becomes murky. Yet something is amiss. The argument proceeds as though the levels of trace gases were solely and directly responsible for a climate shift. What about the Milankovitch mechanism; is that not a necessary trigger? We consider this in detail later on.

But carbon dioxide was not the only trace gas to occupy Hansen's basket. His basket, you recall, contained the full panoply of greenhouse gases. And methane is among the most compelling.

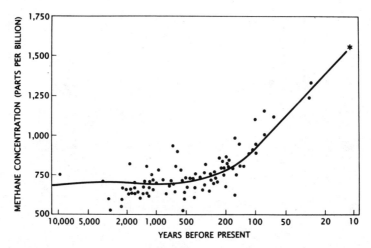

Figure 21. The long, steady level of atmospheric methane at about 750 ppb appears to have begun rising in recent times.

METHANE

Whereas the presence of CO_2 in air became known in the latter part of the 18th century, the presence of methane was unknown until 1948. Like CO_2, it is colorless and odorless and, like CO_2, it is an IR absorber so that it can reasonably be expected to have an impact on climate. From its first analytic measurements in the late 1960s to the most current, it has been found, as shown in Figure 21, to be rising steadily. In fact, it is rising at twice the rate of CO_2. Its current tropospheric concentration is approximately 1.7 ppm (1700 ppb) (0.0002%). Records in the Greenland and Vostok (Antarctic) icecores indicate that since the 17th century (Figure 22), methane (CH_4) levels have more than doubled. Estimates suggest that as much as 50% of this can be attributed to human activity and that the annual 1.5-ppb/year increase is well beyond natural fluctuations.

The human activity in question is primarily biological in origin. Methane is released into the air as a by-product of the anaerobic metabolism of microbes in the gut of such cud-chewing herbivores as cattle, sheep, goats, buffalo, and camel; from decomposing waste in landfills; from water-logged soils such as swamps and bogs; from the digestive tract of termites; and from rice paddies, coal mines, and burning vegetation. The contributing sources, while not endless, are many and diverse and together release from 400 to 700 million tons annually. Table 2.2 lists the sources and the percentages of their contributions. Given the welter of sources, the uncertainty reflected in the wide range of their tonnage is understandable.

It is also fairly evident that these singular sources do not lend themselves to quick fixes—a recent report to the contrary by the U.S. EPA notwithstanding. Bogs, swamps, termites, both wild and domestic ani-

How Has the Atmospheric Concentration of CH₄ Changed?

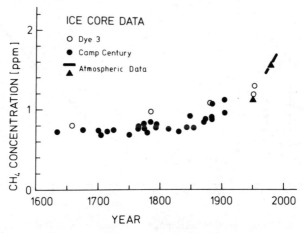

Figure 22. Ice cores obtained at several sites in Greenland allowed determination of methane levels and clearly document its rise over the past 200 years.

mals, oil, and natural gas are indeed natural and not subject to tinkering. Rice paddy acreage continues to grow in extent as the world population increases and, with it, the demand for food. An area of scientific debate and ongoing study is the contribution of termites. Do they produce 5 million tons of CH_4 per year or 150 million? It would be helpful to know, but what could be done should the higher level prove accurate? Livestock is a major contributing source and, as you can imagine, a fine degree of creativity must be brought to bear when estimating the flatulent activity of the various species, as well as obtaining a reliable census of both domestic and wild animals.

Table 2.2
Biological Sources of Methane

Source	Annual Production (10^{12} g CH_4/year)
Enteric fermentation of animals (cows, sheep, horses, goats, buffalo, elephants)	90–130
Rice paddy emissions	70–300
Swamps, marshes	100–200
Biomass burning and wildfires	40–75
Termite emissions	25–125
Freshwater lakes	1–25
Tundra	15–35
Oceans	1–17
Landfills	30–70
Total	372–1000

With the wide disparity in econmic and educational levels around the world, it is difficult to understand the EPA's confidence that a reduction of as much as 50% in methane emissions of ruminant animals could readily be achieved by dietary alterations. Although the sources of methane production appear to suggest areas potentially amenable to control, they may prove highly refractory. One thing is certain: a good deal more information is needed about why methane is increasing as fast as it is, as well as the chemical reactions into which it enters. Given the current scientific effort to understand methane, it is of more than passing interest to note that, as late as 1975, a major scientific report viewed methane quite differently. "To the best of our knowledge," the panel wrote, "most atmospheric CH_4 is produced by microbial activity in soil and swamps under anaerobic conditions . . . . The estimated annual natural production is so large that any anthropogenic sources constitute minor fractions. For this reason, and because CH_4 has no direct effects on climate or the biosphere, it's considered to be of no importance for this report" (12). In fact, methane takes on importance because it is a more efficient absorber of IR radiation then CO_2. On a molecule for molecule basis, it is some 20–50 times more effective, thereby assuming an importance in the greenhouse far beyond its atmospheric level. Consequently, with the ever-increasing world population, shortly to reach 6 billion, there can be little doubt but that CH_4 will also continue to increase, if for no other reason than the growing demand for food and living space. Thus predictions of CH_4 rising to 3.3 ppm by the year 2050 may not be unrealistic. Most recently, relatively large emissions of methane have been identified as coming from another and unusual artificial source. Methane appears to be released into the atmosphere from nuclear power plants fitted with pressurized light water reactors. Prevalent in the northern hemisphere, these have been a new contributor since the 1960s. Release of methane occurs as a consequence of a series of reactions in which hydrogen gas reacts with organic impurities in cooling water. Here then is an opportunity for control measures. It should be a source of concern.

On the other hand, scientists at NOAA's Aeronomy Laboratory in Boulder, Colorado, recently reexamined the rate at which methane is removed from the atmosphere, or more specifically its atmospheric lifetime. Although methane is constantly added to the atmosphere, as noted previously, it is also removed via its reaction with the hydroxyl radical (OH):

(a) $CH_4 + OH \rightarrow CH_3 + H_2O$.
(b) $CH_3 + OH \rightarrow$ products (that do not react with OH).

The hydroxyl radical is often referred to as the "detergent" of the atmosphere, responsible for the removal of many gases produced by natural and human activity. Vaghjiani and Ravishankara measured the rate coefficient in carefully controlled experiments and found that the gas is removed less rapidly than previously assumed. Consequently, methane's lifetime is probably 25% greater than it is currently understood to be. It may also be produced at a slower rate (13). The major effect of this reevaluation will be to revise the methane budget, and modelers will have to revise their equations and recalculate warming estimates, considering that methane is believed responsible for some 12% of the climate forcing of the 1980s.

Given the decreased rate of loss, it may be that atmospheric methane may be stabilized by only a 15–20% reduction in emission. Accordingly, landfills and fossil fuel combustion may be the most appropriate sources of intervention.

CH_4 and CO_2 do not make O_3, but, like methane and carbon dioxide, ozone is a critical component of the stratosphere.

OZONE

Ozone, a form of oxygen in which each molecule contains three atoms of oxygen, exists as an encircling band of gas within the upper reaches of the stratosphere. Here, in a highly chemically reactive area, ozone is constantly destroyed and regenerated. Thus it is unique among the trace greenhouse gases in having neither natural nor artificial reservoirs. It is constantly being produced in a series of photochemically catalyzed reactions in which oxygen molecules (O_2), migrating up from the troposphere, react with ultraviolet light at wavelengths below 0.24 μm and are split into highly reactive oxygen atoms that readily recombine with adjacent oxygen molecules to form ozone—and heat. The reactions proceed as follows:

$$O_2 + (UV) \rightarrow O + O.$$

The energy released in this exothermic reaction is taken up by any number of nearby chemical species X. Thus

$$O + O_2 + X \rightarrow O_3 + X.$$

As X becomes highly energized, it moves faster and, in so doing, becomes warmer. This constantly occurring set of reactions is responsible for a warm stratosphere—where the temperature is higher than at the top of the troposphere. As such, the presence of a warm layer on top of a cooler layer acts as a lid on convection in the troposphere.

With solar radiation most direct and thus strongest above the equator, the photodissociation of oxygen is greatest there. And it is from that region that the newly formed ozone is carried around the earth and across the poles by stratospheric winds. The resulting concentration of ozone, as shown in Table 1, is less than one-millionth that of oxygen. Nevertheless, its trace concentration notwithstanding, ozone is a strong IR absorber, having, as the other trace gases do, multiple spectral bands. As Figure 7 shows, with a major band at 9.6 μm, well inside the otherwise transparent atmospheric "window," its importance as a greenhouse gas is enhanced. However, of even greater importance is its function as a protective shield, blocking the passage of incoming ultraviolet radiation so that little of it reaches the earth. It is, as Figure 7 shows, the only greenhouse gas that also absorbs in the short-wave regions.

Ultraviolet (UV) radiation extends over a range of wavelengths, from the lower end of the visible at 0.4 μm down to the very short at 0.03 μm. On the basis of adverse biological effects, UV has three distinct ranges:

UV-A	0.4–0.329 μm
UV-B	0.330–0.290 μm
UV-C	Below 0.29 μm

Although an intact ozone layer does absorb much of the dangerous UV, a portion of the UV-B does reach the earth's surface and is responsible for photosensitive reactions such as sunburn and tanning—which, in caucasians, as we shall see, does cause cell damage. On the other hand, it may be that people overexpose themselves during hours of the sun's strongest, most direct rays. If time in the sun were reduced, or attire were a bit more protective, UV might prove less of a problem. Nevertheless, the fact that more of the UV doesn't reach sea level is due directly to the fixity and stability of the ozone layer. Without this shield, life on earth would be tenuous at best, given UV's cellular toxicity.

Thus far, UV-A and UV-C have not been a concern to any life forms at sea level. UV-C in the 0.24–0.29-μm range is virtually eliminated by ozone. But because UV-B and UV-C span the photoabsorption spectrum of both DNA and RNA, the major nucleic acid "building blocks" of protein, ozone becomes crucial to the viability of all life. Since UV-C is particularly damaging to DNA, it's the wavelength of choice in germicidal lamps. UV-A has little DNA-damaging impact and thus poses little or no threat. Because the threat of increased UV penetration of the atmosphere is real, we shall discuss its potential adverse human health-related effects in detail later on.

As ozone absorbs so strongly in the infrared and, in turn, reradiates earthward, it has been estimated that as much as 20% of the total energy flux is contributed by ozone. Thus it plays a singular role in atmospheric heat balance.

Given the chemistry of the atmosphere, the level of ozone should remain constant. Not being present as a fixed resource, there is no problem of supply; the stratosphere should not run out of it. Nevertheless, recent data from the southern hemisphere suggest the possibility of a real decline. The observations also indicate that chlorine atoms appear to be involved. But from where would chlorine come? A look back at Table 2.1 shows not a molecule of chlorine in any of the atmospheric gases. As nature abhors chlorine atoms, we must look to the chemical industry.

HOLES AT THE POLE

If any manufactured chemical could be said to be perfect, it is the CFCs, the chlorofluorocarbons. They were literally made to order. In 1928, in response to a need to find a household refrigerant better than ammonia, sulfur dioxide, or propane, one that would be less corrosive, nonexplosive, and nontoxic, chemists at the Dayton, Ohio, Fridigiaire plant (at that time a subsidiary of General Motors) took on the formidable task. In 2 years, they solved the problem and delivered an organic compound containing one or more carbon atoms along with chlorine and fluroine. It was a scientific tour de force—a creative solution to a vexing problem. With such innate characteristics as chemical stability—inertness—nonflammability, and nonexplosivity, they had succeeded beyond expectations. The CFCs were born. The broad uses to which chlorofluorocarbons have been put are shown in Figure 23. They revolutionized household and commercial refrigeration and made possible automobile air conditioning. And with their excellent insulating qualities, they found ready acceptance in the marketplace. The public gobbled up the many CFC-containing products, especially the aerosol spray cans in which chlorofluorocarbons were the propellent solvents. And when CFCs are bubbled into liquid plastic, a foam is produced that can readily be molded into lightweight, highly insulating materials for

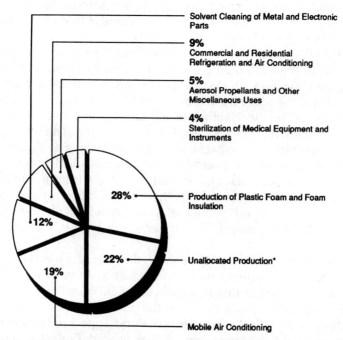

Figure 23. The CFCs have been put to a wide variety of uses. Air conditioning and refrigeration will be difficult to replace. (Courtesy of the U.S. EPA, Washington, DC.)

coffee, soup, ice tea cups, or trays for packaged fresh meats, to say nothing of their excellent fire extinguishing capabilities and their use in pillows and cushions. Their use in refrigerators significantly reduced electric power demand, which meant less combustion of fossil fuel. More could not have been asked of a chemical in the service of humankind. By any set of criteria, they were a success. Until . . . .

In 1971, James Lovelock discovered the presence of minuscule quantities of CCl_3F—trichlorofluoromethane—in the atmosphere over Adrigale, County Cork, Ireland. From the atmospheric point of view, the inertness and relative water insolubility of the dozen or so CFCs, more commonly referred to as Freons, should have been a stunning plus. It turned out to be its Achilles heel. In addition, inherent in the molecule was a fatal flaw (14).

When released or when escaping into the air, these gases pass through the troposphere unscathed, as it were; other less stable chemicals would degrade. Not so the Freons. They continue to rise into the stratosphere. This too should be surprising, since they are far heavier than air. The molecular weight (MW) of trichlorofluoromethane, Freon 11, is 137; air has a MW of 29. Obviously, the wind-driven atmosphere carries molecules of all sizes along with it. The CFCs rise to the level of the ozone layer and above. At the top of the stratosphere, the bottom of the mesosphere, ultraviolet radiation is intense, striking the CFC molecules and splitting them with the release of energetic chlorine atoms—which are free to react with ozone.

In 1975, shortly after their presence in the stratosphere was made known, Rowland and Molina, of the University of California at Irvine, published their seminal paper presenting a theoretical chemical model for the ozone-destroying effects of the CFCs (15). Given their chemical inertness, the CFCs and ozone do not interact. The problem occurs only as a consequence of the CFCs coming in contact with ultraviolet radiation. The mechanism offered by Rowland and Molina proceeded from the photolytic dissociation of chlorine-containing carbon compounds with the subsequent liberation of chlorine, which in turn attacked ozone (O_3), liberating an atom of oxygen. The result is loss of ozone. With impaction of short-wave ultraviolet, each CFC molecule splits and releases atomic chlorine, which then attacks adjacent ozone molecules:

$$(1) \quad CFCl_3 + UV \rightarrow CFCl_2 + Cl.$$
$$(2) \quad Cl + O_3 \rightarrow ClO + O_2.$$
$$(3) \quad ClO + O \rightarrow Cl + O_2.$$

Reactions (2) and (3) continue almost indefinitely, so that one chlorine atom can destroy 100,000 ozone molecules. Trichlorofluoromethane is photolyzed to $CFCl_2$ and free radical atomic chlorine, which reacts with ozone to form chlorine monoxide and oxygen, effectively destroying ozone. The chlorine monoxide in reaction (3) interacts with atomic oxygen to produce oxygen and chlorine. This was an elegant schema, but there was little to confirm it. Not for a decade.

In 1985, scientists of the British Antarctic Survey found that springtime levels of ozone over Halley Bay had decreased by 40–50% between 1977 and 1984 (16). The normal level of 300 Dobson units had declined to 180. Data from NASA's *Nimbus 7* orbiting satellite (Figure 24), with its Total Ozone Mapping Spectrophotometer (TOMS), confirmed the loss as well as the presence of a "hole" in the ozone layer whose width encompassed the entire continent. Here then was confirmation of the Rowland–Molina hypothesis. Figure 25 is a reproduction of the original TOMS data, for October 5, 1987, showing the mean ozone levels in Dobson units.* Figure 26 compares the years 1986, 1987, 1988, and 1989. The lowest levels were attained in 1987, while the even years, 1986 and 1988, had far less reductions than the odd years. But the truly sublime relationship between stratospheric ozone and chlorine dioxide levels, especially that on September 21, 1987, is shown in Figure 27. The September trend lines exhibit the type of relationship for which researchers hunger in their search for convincing cause–effect relationships. Here we see that, in the 30 days between the two flights, 80% of the ozone initially present within the vortex disappeared. It was within those 30 days that winter darkness dissolved into austral spring, fetching with it the chemically catalyzing effects of sunlight. From Figure 28, the 30-year declining trend in DU values at three locations over Antarctica is evident. The magnitude of Antarctica compared to the continental United States is shown in Figure 29, while Figure 30 depicts the area encompassed by the "hole"—several million square miles. By way of comparison, the area of the 48 contiguous United States is 2,962,030 square miles. Small amounts of chlorine can readily destroy large amounts of ozone.

*The Dobson spectrophotometer measures the relative intensity of transmitted light at two distinct UV wavelengths—one strongly and one weakly absorbing—to determine the total amount of ozone. One DU represents a single molecule of ozone per billion molecules of air: 100 DU is the equivalent of a layer of ozone 1 millimeter thick around the world.

Figure 24. (*a*) Illustration of the *Nimbus 7* satellite in space. (*b*) The *Nimbus 7* spacecraft components, consisting of the integrated subsystems that provide the power, attitude control, and information flow required to support the payload for a period of 1 year in orbit, are contained within the three major structures of the spacecraft. The three consist of a hollow torus-shaped sensor mount, the solar paddles, and a control housing unit that is connected to the sensor mount by a tripod truss structure. The spacecraft weighs 965 kilograms and has a configuration similar to that of an ocean buoy. It is 3.04 meters tall, 1.52 meters in diameter at the base, and 3.96 meters wide with the solar paddles fully extended. The sensor mount that forms the satellite base houses the electronics equipment and battery modules. The lower surface of the torus provides mounting space for sensors and antennas. A box-beam structure mounted within the center of the torus provides support for the larger sensor experiments. The control housing unit is located on the top of the spacecraft and above this unit are the sun sensors, horizon scanners, and a command antenna. (Courtesy of GE Astro-Space Division, Princeton, NJ.)

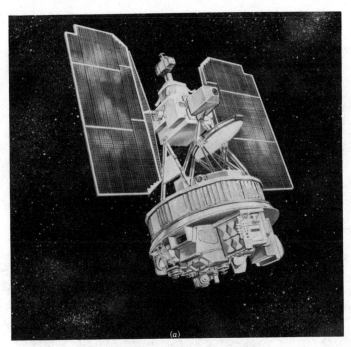

(a)

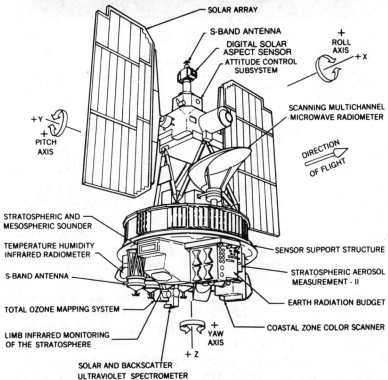

SOLAR ARRAY

S-BAND ANTENNA

DIGITAL SOLAR
ASPECT SENSOR

ATTITUDE CONTROL
SUBSYSTEM

+
ROLL
AXIS

+X

SCANNING MULTICHANNEL
MICROWAVE RADIOMETER

+Y

+
PITCH
AXIS

DIRECTION
OF FLIGHT

STRATOSPHERIC AND
MESOSPHERIC SOUNDER

TEMPERATURE HUMIDITY
INFRARED RADIOMETER

S-BAND ANTENNA

SENSOR SUPPORT STRUCTURE

STRATOSPHERIC AEROSOL
MEASUREMENT - II

TOTAL OZONE MAPPING SYSTEM

EARTH RADIATION BUDGET

LIMB INFRARED MONITORING
OF THE STRATOSPHERE

COASTAL ZONE COLOR SCANNER

+
YAW
AXIS

+Z

SOLAR AND BACKSCATTER
ULTRAVIOLET SPECTROMETER

Nimbus 7 Observatory

(b)

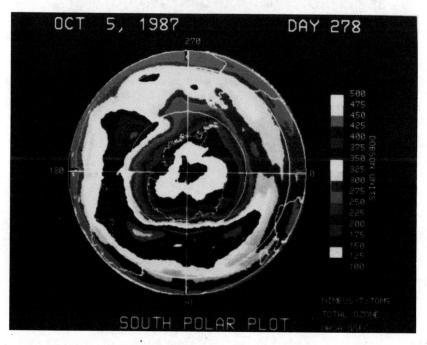

Figure 25. This is a southern hemisphere plot of total ozone distribution for October 5, 1987. The region bounded in purple is of reduced ozone—less than 200 Dobson units. Within the purple area is a black region where the ozone is less than 125 DU. The lowest level ever observed, 109 DU, is located in this black area near the South Pole. This plot also shows that the ozone hole area is larger than the Antarctic continent. (Courtesy of NASA, Greenbelt, MD.) (See color section.)

This "hole" was not only troublesome to scientists, but its heavy media coverage produced worldwide alarm that a thinner or destroyed ozone shield would permit biologically harmful UV-B radiation to reach the earth, increasing the potential for skin cancer, cataracts, and other health-related abnormalities. A hole over the south Pacific, south Atlantic, and Indian Oceans could be dangerous to marine life, the food chain, and global warming. Large-scale destruction of plankton, for example, the microscopic drifting plants and animals inhabiting the upper layers of seas and oceans—which normally remove great quantities of CO_2 from the oceans by incorporating it into their carbonate-containing shells—would make this CO_2 available to further increase the levels of greenhouse CO_2 and thereby add to global warming; a pernicious example of the unbalancing of the global carbon cycle. If, with continuing destruction, it opened even wider, it could place the populations of large areas of New Zealand, Australia, and South America at risk. If a hole opened over the Arctic, the people of Sweden, Norway, Denmark, Finland, and perhaps Scotland could be jeopardized. Thus far, large seasonal ozone depletion

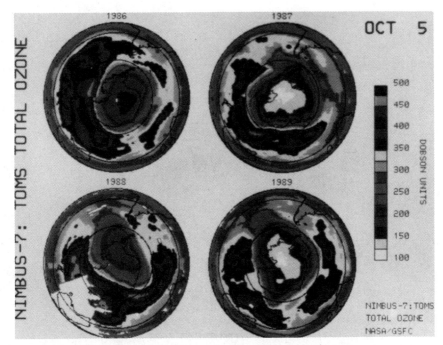

Figure 26. These southern hemisphere plots show the total ozone distribution for October 5, 1986, 1987, 1988, and 1989. The depth of the ozone hole has been following an approximately 2-year cycle; relatively deep in odd years and not as deep in the even years. (Courtesy of NASA, Greenbelt, MD.) (See color section.)

appears limited to the south polar region. The sketch in Figure 31 looking up from below the Antarctic shows an area of ozone thinning and its relation to the polar vortex. How does this thinning occur, and can it be prevented or repaired?

During the long sunless Antarctic winter (March to August) ozone is present at levels considered normal in the 1960s and 1970s. During the austral winter, the air over the continent is, in a sense, walled off from the rest of the atmosphere by the swirling polar vortex of subzero winds, which sweep around the Antarctic. These circumpolar winds are so cold that the air temperature falls to -90 to -100°C, cold enough for what little water vapor present to freeze into crystalline clouds. Chemical reactions, even at these temperatures and blackness of the polar night, can apparently occur on the surface of these water (cloud) crystals. Thus with the coming of spring (September to October) and the appearance of the sun, the chlorine molecules are photolyzed to atomic chlorine, which participates in a series of not yet fully understood chemical reactions that ultimately destroy ozone. Most importantly, as indicated, chlorine is little affected in the process. Accordingly, each chlorine atom can destroy thousands of ozone molecules before it is inactivated or carried down to the tropo-

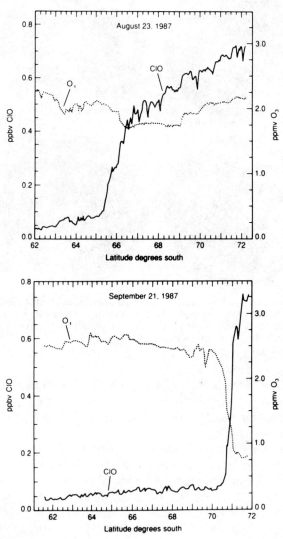

Figure 27. Chlorine oxide and ozone concentrations over Antaractica at 18-km altitude, August 23 and September 21, 1987, as measured on ER-2 aircraft. (Reproduced with permission from *Environmental Science and Technology*, American Chemical Society, Washington, DC.)

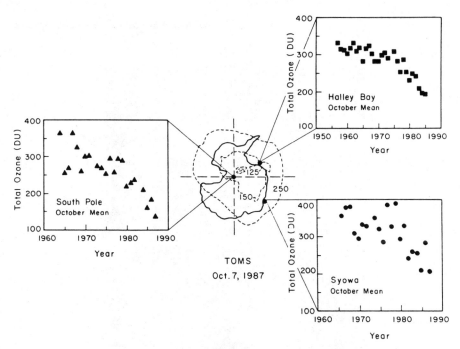

Figure 28. Observational data that first indicated the existence of the Antarctic ozone hole. (Reproduced with permission of the National Academy Press, Washington, DC.)

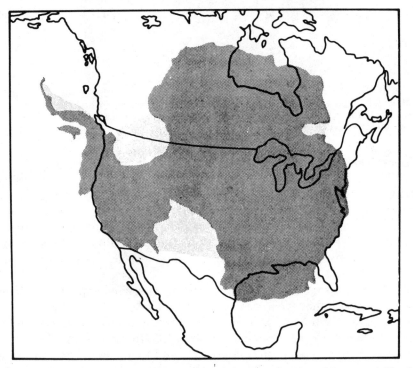

Figure 29. Size of the Antarctic continent (gray) relative to the United States and Canada. Clearly, Antarctica is larger than the continental United States.

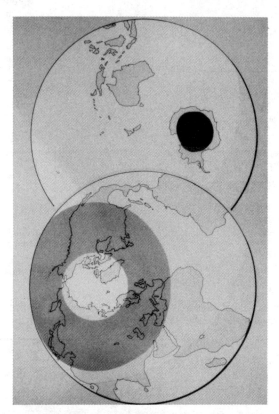

Figure 30. The size of the ozone hole on October 5, 1987 was about as large as the entire continent. If an area of similar size opened over the northern hemisphere it would place at risk a large proportion of the population of Europe, North America, the Middle East, India, China, and the Soviet Union.

sphere. These reactions continue for almost 2 months throughout the spring, causing the steady depletion of ozone and the resultant "hole." As spring moves into summer, warmer air dilutes and dissolves the vortex and stratospheric ozone levels return to normal—via the process noted earlier.

By September 1987, the "hole" was the deepest and widest seen thus far; and in September 1989, it was observed yet again and tests demonstrated that the continuing use of CFCs had increased its atmospheric level by an additional 10%.

Declines in ozone levels in the warmer northern hemisphere have been found, but not at the levels seen at the far colder South Pole. The reason for this appears to be the unique meteorological conditions over the Antarctic, which influence ozone decline. Polar stratospheric cloud formation within the circumpolar vortex is one of the conditions that does not occur over the warmer Arctic. Another is the fact that warming occurs after the arrival of sunlight and, with it, ozone reduction. In the Arctic, warming occurs

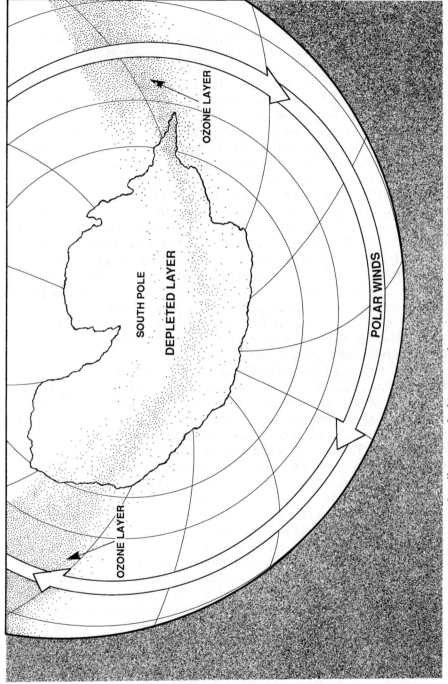

Figure 31. In this illustration, looking up at the Antarctic continent from below, we see an area of reduced ozone within the swirling polar vortex.

before the arrival of sunlight, which is necessary for driving the chemical reactions. These different conditions may well account for the large differences in ozone depletion between the poles. In June 1990, the British Antarctic Survey reported a 6% reduction in the ozone layer in the high northern latitudes in winter, which has intensified efforts by Sweden, Norway, and Finland, along with Australia and New Zealand, to achieve a ban on CFCs. However, according to Michael Proffet of NOAA's Aeronomy Laboratory, the major ozone reductions observed over the South Pole should not be expected over the Arctic because of the very different meteorological conditions. The ozone depletions over the North Pole are, in no sense, the hole seen over the Antarctic. As for the South Pole, Michael J. Prather and Robert J. Watson of NASA recently calculated that the hole over that pole could be repaired by the year 2100 if, and perhaps only if, there is a worldwide phaseout of all CFCs and other halocarbons by the year 2010 (17). Their calculations showed that the total stratospheric level of chlorine-containing compounds was about 3 parts per billion and needed to be reduced below 2 ppb, if the hole were to attain its normal density. Each year's delay, they predicted, pushes the 2-ppb date 3.6 years into the future. Furthermore, it takes 8–10 years for CFC molecules to pass through to the top of the stratosphere. Thus abruptly halting their use cannot be expected to produce benefits for decades. Nevertheless, the call for an outright ban is soundly based.

Indeed it was. Early in April 1991, to worldwide consternation, the U.S. EPA announced that the ozone shield had thinned another 4.5–5%. That conclusion was based on direct satellite measurements. But, unlike previous declines, this one was not limited to the polar regions; this was over the continental United States. This was a new and wholly unexpected development. The cause of this most recent thinning is unknown, but the U.S. EPA has instituted a new study to attempt to fill the void. Meanwhile, it has been suggested that the growing number of solid fuel rockets lofted into space may be the culprit. Apparently, ozone is destroyed by rocket propellants, which inject tons of chlorine into the stratosphere. But this too remains highly controversial and must be resolved.

That the ozone layer remains under assault is shown in Figure 32. Here we see that although the Montreal Protocol required the 57 signatory nations to freeze production and use of the five major CFCs at 1986 levels, followed by phased-in reductions of up to 50% by 1999, those reductions would continue the upward trend of stratospheric chlorine levels. When the total phase-out adopted in June 1990, under the London modification of the Protocol, goes into effect—that is, the developed nations totally phase out CFCs, while the less developed countries phase them out by 2010—chlorine levels are calculated to peak by the year 2005, followed by a steady decline. By 2035, the ozone level should be no less than that detected in 1986. At that time, the ozone level should be at full strength once again. This scenario is predicated on the hope that the nations of the world keep the pledges that were made in London. One of those pledges

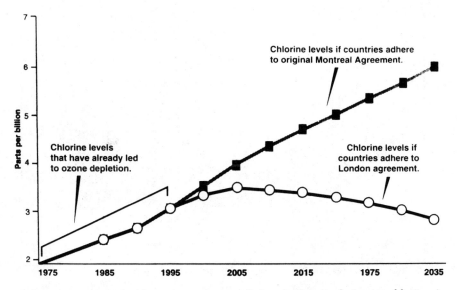

Figure 32. The ozone shield remains under assault. Levels of ozone-destroying chlorines in the atmosphere would continue to rise under international controls adopted in Montreal in 1987, allowing more solar radiation to penetrate the weakened shield. Under more stringent controls adopted in London in 1990, concentrations are expected to peak about 2005 and then decline. (Courtesy of U.S. EPA, Washington, DC.)

called for the United States to provide 25 million dollars of financial aid to the less developed countries to assist in their switching to more ozone-friendly chemicals. According to Eileen Clausen of the U.S EPA, the funds are already in her budget and she foresees no hitches, because switching is in everyone's best interests.

In response to questions about how much more damage to the ozone layer could be expected, an EPA ozone specialist was quoted in *The New York Times* to the effect that "we really don't know. Earlier attempts to forecast ozone depletion based on mathematical models were proved so inaccurate that basically we have thrown out the models" (18). What is one to think? If we cannot rely on models, what hope is there? Models have also predicted additional skin cancers as a result of thinning of the ozone layer. This aspect of the problem is discussed in Chapter 4.

A FATAL FLAW

The fatal flaw remains. Inherent in the CFC molecule is infrared absorbency. Not only is it a strong absorber, making it a new and significant greenhouse gas, but it absorbs at a most unusual spectral region. As shown in Figure 33, many of these CFCs array themselves in the long-wave region

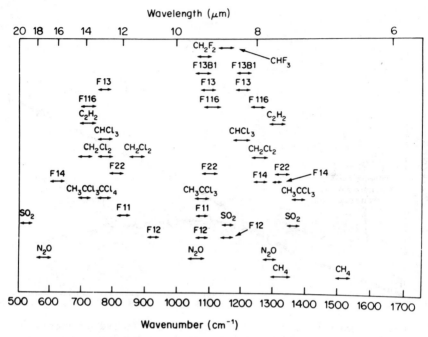

Figure 33. The arraying of CFC molecules in the area of the atmospheric "window" blocks the "window" and accounts for their greater efficiency of IR absorbance.

between 7 and 13 μm, the area of the transparent "atmospheric" window through which radiation escapes back to space. Collecting here, as they do, virtually blocks the "window," in effect closing the atmosphere's "relief value." Thus not only do the CFCs create additional trouble in the greenhouse, but they do so far beyond their ppb levels. On a molecule for molecule basis, they are 4000 times (trichlorofluoromethane, Freon 11) and 10,000 times (dichlorodifluoromethane, Freon 12) more energy absorbing than a molecule of CO_2. Why this occurs is of more than passing interest. As with all trace gases, thermal infrared radiation is absorbed and induces vibrations in the bonds holding the atoms in a molecule together. Carbon dioxide, for example, actually has a carbon atom between two atoms of oxygen: O=C=O. Upon absorbing thermal energy, the molecule vibrates at discrete frequencies corresponding to the frequencies of infrared radiation. CO_2, as with all trace gases, has multiple absorption bands. One at 4 μm is not an energy active band but, at 15 μm, vibrational activity is high. The CFCs strong vibrational excitement occurs between 7 and 13 μm, and gases that absorb in the window region, as the CFCs do, block radiation from the warmer troposphere and as a consequence produce a greater warming effect than a gas that absorbs at 15 μm.

Hansen recently noted that "the increase of CFC forcing in the 1980s represents about one-quarter of the total growth in radiative forcing by

a

b

Figure 1.13a, b. (*a*) A typical length of ice core, while relatively heavy, can be held in the palm of the hands and has great clarity. (Courtesy of Dr. Bruce Koci, Polar Ice Coring Office, University of Alaska, Fairbanks.) (*b*) Bubbles of air trapped in Antarctic ice. (Courtesy of Claude Lorius, Laboratory of Glaciology and Geophysics of the Environment. St. Martin d'Heres, France.)

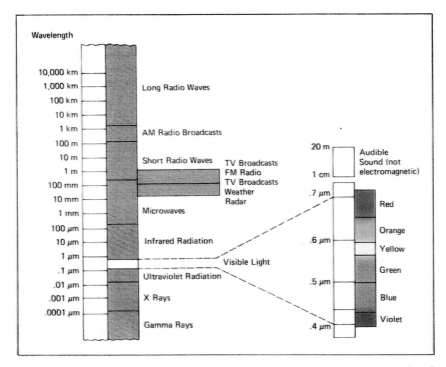

Figure 2.2. Profile of the electromagnetic spectrum portraying the great range of wavelengths and the fact that the visible band is only a small fraction of the total.

Figure 2.5. This photograph, a cross section through the Earth's atmosphere over the Pacific Ocean, was taken from the Space Shuttle *Challenger*, on October 10, 1984, from an altitude of 123 nautical miles. Stratosphere appears blue, while the troposphere appears white and pink. (Courtesy of M. R. Helfert, NASA — Johnson Space Center, Houston, Texas.)

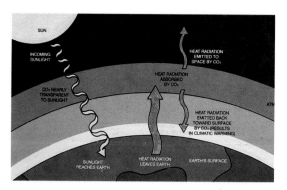

Figure 2.8. Diagrammatic representation of the radiation exchanges — absorption and reflection — which result in the beneficent "greenhouse" effect.

Figure 2.11. Oxygen-rich bubbles on the leaves of submerged green plant show that photosynthesis is taking place. (Courtesy of Quesada/Burke, New York.)

Figure 2.13. Site of the Scripps/NOAA Laboratory 3400 meters above sea level on barren Mauna Loa. In addition to measuring CO_2 levels, measurements of aerosols, ozone, and solar coronas are also made here. (Courtesy of C. D. Keeling, Scripps Institution of Oceanography, La Jolla, CA.)

Figure 2.25. This is a southern hemisphere plot of total ozone distribution for October 5, 1987. The region bounded in purple is of reduced ozone — less than 200 Dobson units. Within the purple area is a black region where the ozone is less than 125 DU. The lowest level ever observed, 109 DU, is located in this black area near the South Pole. This plot also shows that the ozone hole area is larger than the Antarctic continent. (Courtesy of NASA, Greenbelt, MD.)

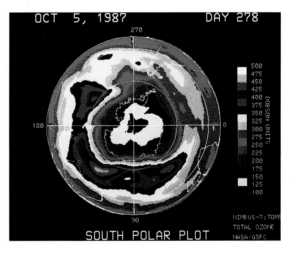

Figure 2.26. These southern hemisphere plots show the total ozone distribution for October 5, 1986, 1987, 1988, and 1989. The depth of the ozone hole has been following an approximately 2-year cycle; relatively deep in odd years and not as deep in the even years. (Courtesy of NASA, Greenbelt, MD.)

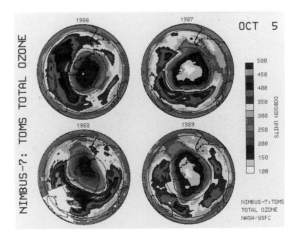

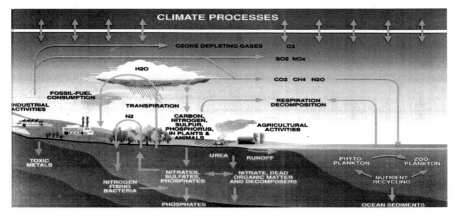

Figure 3.2. Climatic processes as currently perceived. (From U.S. Department of Justice, Report of the Task Force on the Comprehensive Approach to Climatic Change.)

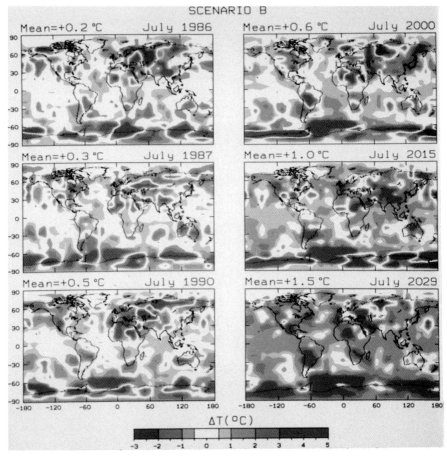

Figure 3.4. Simulated July temperature anomalies for the 6 years indicated. The simulation is based on a specific set of boundary conditions. (Courtesy of James E. Hansen, Goddard Institute for Space Studies, New York.)

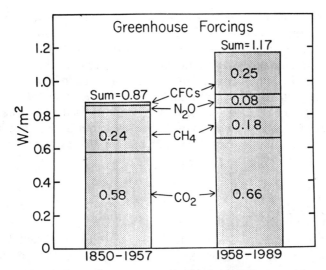

Figure 34. While the bars show the increased greenhouse forcing of several naturally occurring atmospheric gases, it is the untoward increase in CFCs that is of foremost concern. (Courtesy of James E. Hansen, Goddard Institute for Space Studies, New York.)

trace gases." Figure 34 shows this relationship in terms of increased watts per square meter (W/m^2). For the period prior to the initial collection of CO_2 levels, greenhouse forcing was at a level of less than 1 W/m^2. Over the ensuing 30 years, an additional watt has been added. As the figure shows, the Freons have contributed substantially to this increased warmth. Perhaps as disturbing is the fact that, even if their production and release stopped abruptly tomorrow, their atmospheric concentrations would continue to increase because of their inordinate atmospheric residence times—on the other of 60–100 years—a consequence of their otherwise superlative chemical nature: inertness and water insolubility. Thus projecting the year 2100 for repair of the "hole" is understandable. In retrospect, it may be well to recall the remarks made by Lovelock, Maggs, and Wade in their article in *Nature* in 1973. "The presence of these compounds," they said (referring to their discovery of the CFCs in the atmosphere), "constitutes no *conceivable hazard*: indeed the interest lies in their potential usefulness as inert tracers for the study of mass transfer processes in the atmosphere and *oceans*." They continued: "The voyage of the R.R.S. *Shackelton* from the United Kingdom to Antarctica and back, in 1971-72, provided an excellent opportunity to compare the observed global distribution of CCl_3F with the predicted distribution from models of the behavior of an *ideal* inert gas" (19). This should stand as a bright, shining example for those individuals who believe that nothing should be introduced into the environment until it is fully tested. Indeed, can anything ever be fully tested prior to use?

Lovelock et al. also had this to say: "The release of CO_2 by combustion has been described as an unscheduled geophysical experiment; if so, it is a rash undisciplined one, because emissions are small compared with those of the large complex natural cycle. By contrast, the emissions of CCl_3F, also unplanned, are much closer to an ideal experiment on a global scale. There are no natural sources; it is chemically and physically inert; *it does not disturb the environment* [italics introduced], and it can be measured accurately to one part in 10^{12} by volume." This last criterion is of some importance, since 10 with 12 zeros after it is 10 trillion! As the CFCs exist in the atmosphere at levels of 1 part in 10^{10}, and parts per billion and less, they might never have been found were it not for the exquisitely sensitive analytic procedure then available.

NITROUS OXIDE

As a consequence of its absorption of long-wave infrared radiation in the 7.8- and 17-μm bands, nitrous oxide can have a direct effect on the earth's climate; and increased emission should increase the greenhouse effect. From Table 1 we see that the present atmospheric concentration of N_2O is 0.3 parts per million. Measurements made between 1970 and 1980 indicate that this natural greenhouse gas has been increasing by 0.25% per year. Preindustrial levels had been estimated at 0.28 ppm.

Nitrous oxide emissions result primarily from bacterial denitrification in the soil. Apparently, species of the bacterium *Nitrobacter* convert nitrates (NO_3) in the soil to molecular nitrogen (N_2), which rises into the atmosphere, where it reacts with oxygen to form N_2O. This is the largest source of atmospheric N_2O and is unrelated to human activity. Estimates also suggest that a doubling of N_2O could increase average surface temperature by as much as 0.3–0.4°C. Since 1981, when Weiss reported that the measured increases in nitrous oxide may be primarily due to combustion of fossil fuel and biomass burning and that production of N_2O from denitrification of nitrogen-containing fertilizers was considerably less than previously estimated, this has been a source of contention because the supporting evidence was severely limited (20). Most recently, however, it has been shown that biomass burning probably contributes far less than heretofore believed. In fact, the new data indicate that biomass burning may contribute less than 7% of atmospheric N_2O, as opposed to earlier estimates of several times that amount. The investigators suggest that estimates of N_2O production from fossil fuel must also be downgraded and that if their data are correct, there is then a significant imbalance in the global N_2O budget. They inform us that "it is possible that a significant global source of N_2O has yet to be identified, or that known sources may have been underestimated" (21). This may have been a particularly prescient remark. Recently, two scientists at the University of California, San Diego, reported that N_2O is generated, but more importantly emitted to the atmosphere, during the manufacture of nylon and may account for as much as 10% of the total increase in N_2O (22). In simulating the industrial production of nylon, they found that N_2O was a by-product of the production of adipic acid, the monomeric dicarboxy acid that forms the nylon polymer. They estimated that worldwide 4.5 billion pounds of nylon are manufactured annually and that a similar amount of N_2O is pumped into the atmosphere. Once again we find humans producing a highly useful and desirable product—along with a not-so-useful one. Given N_2O's long residence time in the atmosphere, from 100 to 150 years, such discharges must be curtailed, and very likely can. Although as Table 2.1 shows, CO_2 is present in the atmosphere at levels 1000 times greater than N_2O, on a molecule for molecule basis, N_2O is about 200 times as efficient an IR absorber and is only slightly less destructive of ozone than the CFCs. Nitrous oxide reacts with oxygen to form nitrogen oxides as follows:

(a) $N_2O + O \rightarrow 2NO$.
(b) $NO + O_3 \rightarrow NO_2 + O_2$.

Thus increased NO levels due to increases in N_2O are destructive of ozone. The conclusion that can be drawn here is that natural denitrification cannot be prevented, nor does it need to be; but anthropogenic emissions can and should be. Nitrous oxide can be brought to stable levels.

WATER VAPOR

Water vapor, our fifth major trace gas, was in fact the first to be so considered. In his series of ingenious experiments conducted in 1860, John Tyndall, a British physicist, found that water vapor had great absorptive capacity for solar radiation. Giving a value of 1 to dried and purified air, he calculated the absorption of ambient air containing water vapor and carbon dioxide as 15. After deducting the effect of CO_2, his figure showed the absorption of water vapor to be 13 times greater than air. He was quick to grasp the implication for climate. In 1861, in his published work, he remarked that "this vapor, which exercises such a destructive action on the obscure [heat] rays is comparatively transparent to the rays of light. Hence, the differential action of the head coming from the sun to earth, and that radiated from the earth to space, is vastly augmented by the aqueous vapor of the atmosphere." While not referring to it by that name, he had uncovered the greenhouse effect some 130 years ago.

A century later, Fritz Möller, of the University of Munich's Meteorological Institute, advanced the provocative concept that water vapor might also act as an amplifying or intensifying positive feedback mechanism (23). Thus, if increases in CO_2 were to induce a warming effect, increased evaporation from the seas and oceans would, according to his mathematical models, increase atmospheric water vapor, which in turn would absorb more long-wave radiation and drive temperatures even higher. Testing this hypothesis had to await 1989 and the clever field trials undertaken by Raval and Ramanathan of the University of Chicago's Department of Geophysical Sciences. They studied the oceans, where water vapor feedback should be most intense. They obtained temperature readings of specific areas of the oceans from ships and buoys around the world. From this, they were able to calculate the heat emitted. By using satellite-collected data on the amount of humidity escaping into space, they were able to infer the amount blocked by a greenhouse effect. According to their estimates, the greenhouse effect increased significantly with sea surface temperature. This finding verified a hypothesis of immense potential impact on climate change and forced a change of view among those who believed that water vapor would act as a negative feedback system if temperature increased; water vapor, it was suggested, would diminish, rather than amplify, a global warming. According to this view, an initial warming would increase the atmospheric content, which would lead to a cooling effect and lessen the greenhouse warming. Raval's and Ramanathan's work scotched this concept and lent greater credence to climate models. "The rate of increase," they reported, "gives compelling evidence for the positive feedback between surface temperature, water vapor and the greenhouse effect; the magnitude of the feedback is consistent with that predicted by climate models" (24). Furthermore, this study proved that orbiting satellites were an effective means for directly monitoring changes in greenhouse effect and would do so in the future. Clearly, the *Nimbus* satellite program, in this instance the Earth Radiation Budget Program (ERBP), has done yeoman service for the country and the world. An essential lesson to be drawn from our increasing understanding of the radiative effects of trace gases is that water vapor, methane, and ozone can be at least as important to climate forcing as CO_2. This must affect the way the problem of excessive heating with the greenhouse is approached. In addition, it has now been estimated that jet airplane flights at altitudes upward of 7 miles produces as much water vapor in a year as the total occurring naturally. For each kilogram (kg) of gasoline consumed (2.2 lb), 1.25 kg of water vapor are produced. With the large and growing volume of airline traffic, this introduces yet another area of uncertainty to model estimates of global warming.

DEFORESTATION

The contribution of CO_2 to the greenhouse burden from the burning of coal and oil is fairly well in hand; estimates for the decades of the 1980s suggest 5–5.5 billion tons annually. But what about emissions from destruction of the tropical forests, which stretch in a belt along the equator from Brazil through Zaire to Malaysia and Indonesia? Here are the earth's great carbon dioxide storehouses—a vast potential for additions to an already over-taxed greenhouse. The areas involved are immense, and the scale of the devastation may be incomprehensible. What does it mean, for example, to say that tens of thousands of square miles of forest are disappearing annually? Several examples may assist comprehension.

According to a recent study by the World Bank, the Lacondona Forest in the Mexican State of Chiapas, the largest tropical rain forest in North America, is rapidly disappearing. Lacondona is the size of Connecticut, some 5000 square miles. Since 1970, 60% of the forest has been lost. The increasing rate and size of loss are readily seen in the 1500 square miles, larger than the entire city of Providence, Rhode Island, lost between 1980 and 1989. And in Peru, to accommodate the Western world's insatiable demand for cocaine, farmers have cleared more than 500,000 acres of virgin forest—780 square miles, 12 times the size of Washington, DC, to plant coco leaves. Half a million acres are also the equivalent of the acreage burned due to natural lightning storms in Yellowstone Park, in 1988. In the state of Rondonia, Brazil, an area 2½ times the size of Connecticut was burned for forest clearing in 1988. Figure 35, a view from space taken by

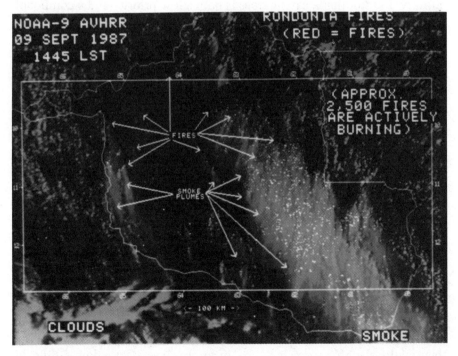

Figure 35. This satellite image was generated as it passed over the state of Rondonia, Brazil. Thousands of fires and smoke plumes are evident. (Courtesy of NASA–Goddard Space Flight Center, Greenbelt, MD.)

the NOAA-9 weather satellite, shows the extent of the fires and smoke. In India, estimates by their National Remote Sensing Center indicate a deforestation rate for the 1980s at almost 4 million acres per year—equivalent to the loss of the entire city of Honolulu. In the Ivory Coast, forest cover has decreased by 75% over the past three decades. An estimated 200 million cubic meters of timber has been burned—not even made available for building use, which would have contained the stored CO_2—to clear the land for agriculture. And in Ghana, if the figures can be believed, 80% of the forests have been leveled.

The rape of Guatamala's lush Peten Forest, a 13,300 square mile area of prize cedar and mahogany, much of it over 100 years old, is a cautionary tale of unbridled greed coupled with laws promulgated to pacify the world-at-large but, in reality, meant to be ineffective. Almost half of the acreage has been destroyed by mindless lumberjacking and clearing for corn. Bootlegging trees, which fetch $2000–2500 per large tree, has drawn an army of drifters, each with a chain saw. Trees are felled like matchsticks and hauled quickly over the border into Mexico, where lumber mills have been set up just for this trade. Guards were indeed hired to enforce existing laws, but they are unarmed and can neither confiscate contraband lumber nor prevent cutting in the first place. The only authority in the region is the military. But, for them, saving trees is both a low priority and a conflict of interest, as it is an open secret that army officers control logging and milling. Thus corruption, violence, and bureaucratic ineptitude prevail. With the forests at the mercy of human predators, it is foolhardy to expect them to remain.

The Burmese (Myanmar) government is ravaging its own forests to pay for guns and planes needed to fight border tribesmen. Satellite photographs have documented the annual disappearance of 1.2 million acres of Myanmar's forests. And with the help of Thai business interests, some 25 million cubic feet of teak flow out of the country to Japan, Hong Kong, Europe, and the United States to be made into furniture. The selection system for removing trees worked out by British colonialists has been replaced by the uncontrolled use of chainsaws, bulldozers, and tractors.

Is any forest immune? In the states of Washington and Oregon, a race is on to log the trees before public pressure and state regulations curtail further removal. Is short-term gain inherent in the human genome? A glimmer of hope may come from Brazil. Concerned Brazilians, including industrialists from the pulp and paper industry, as well as scientists from the University of Saõ Paulo, recently enunciated an ambitious plan to reforest large areas of the country. The Floram Project (Florestas para o ambiento—Forests for the Environment) envisions planting 10 billion trees over 201,000 square kilometeres (some 21 million acres), claiming they could absorb as much as 5% of atmospheric CO_2, thus aiding in offsetting the greenhouse effect. Although José Goldemberg, Minister for Science and Technology, is one of Floram's backers, others are less sanguine, believing that the 20 billion dollars required will not materialize from countries at large. However, there is expectation that the various Brazilian states will adopt the project for their areas, in conjunction with local industry. The goals are admirable: fight global warming, improve local conditions, prevent soil erosion, provide raw material for industrial use, and prevent further destruction of existing forests by providing new sources of wood for industrial use. However admirable the goals, the project may never get beyond the planning stage if funds to support the planting fail to appear.

Recently, hope for the future of Lacondona arose. An accord has been negotiated and agreed upon between the Mexican government and Conservation International, a private American ecological group. They have entered into a "debt for nature" exchange in which Mexico has agreed to accept a $4 million reduction of its foreign debt in return for specific commitments aiding the preservation of the forest. Conservation International will assume the $4 million debt. Because of discounting and financial restructuring, it will actually cost less than $2 million, in return for guarantees that, before 1994, the government will finance a number of needed projects, such as a scientific field station and a conservation data center at the National Autonomous University of Mexico. While this is a precedent-setting development, it does raise a disturbing question: Will governments drag their feet on forest protection, waiting for external aid to cover their balance of payments? In the process, huge acreage could be wiped out. On the other hand, it just may be an idea whose time is at hand.

Logging for the timber market, clearing land for farm and pasture development as well as urban growth, and flooding for hydroelectric power plants for the production of electricity are all well established procedures—all reasonable activities on the face of it. More than likely, little attention would be given to these many loses if it were not for concern over the release of CO_2, which, estimates suggest, ranges from 0.4 to 2.5 billion tons per year. Recall too that, with the loss of trees and plants, there is a loss of the photosynthetic process, which removes huge amounts of CO_2 from the atmosphere. Thus the loss of trees, particularly with their burning, adds additional CO_2 to the atmosphere, but the loss of a CO_2 sink causes additional atmospheric accumulation of CO_2. Furthermore, tremendous amounts of moisture are generated by the forest flora; clouds are developed in large numbers, which play a major role in the earth's total radiation budget. With forests gone, so are moisture and clouds. Over large areas, this loss of moisture could mean adding additional weeks or months to seasons already dry. Thus the short-term gain will prove doubly harmful.

Additionally, suggestions that replacing old forests with new ones could help slow the addition of CO_2 have not been borne out. Recent studies have shown that old forests store huge amounts of carbon and, when they are cut, the carbon can remain in storage in the timber used for building or be released during production of paper, wood burning, and decay. Whereas young trees have higher primary productivity and remove carbon more rapidly than old ones, the net effect is to add CO_2 to the atmosphere. Here is an excellent example of the need for firm data to aid in decisionmaking. This brings us back to the broad estimate of 0.4–2.5 billion tons per year as the contribution of deforestation to atmospheric CO_2. If it is closer to 0.4, that is one thing; if 2.5, quite another. Then again, perhaps it is neither. Both the reporting and data gathering leave much to be desired. Does it matter? That is the central issue. Are the levels estimated to be added over the next 20–30 years sufficient to produce the predicted climate shifts or would additional emission for deforestation shorten the time or heighten the effects or, again, would it not matter because the carbon cycle is so vast and accommodating that any level of anthropogenic increase can readily be integrated and balanced? Are answers to these question necessary? If so, how will they be obtained. And when?

FIRE

Between April and June of 1987, two separate fires raged on either side of the Amur River. Begun within 2 weeks of one another, these infernos are now considered the worst fires the world has known in 300 years. On the Chinese side, the area of the Great Hinggan Forest stretches some 400 miles along the Sino–Soviet border in China's most northern Heilongjiang province, where the Amur River, or Black Dragon as the Chinese call it, separates China from Russia.

On the Chinese side, the fire's immediate cause was carelessness on the part of an inexperienced worker, who spilled gasoline from a power saw and then tried to light a cigarette: in a dry forest with high winds, the scenario was set. The fire burned for 32 days and in that time the world's largest stand of virgin conifers was destroyed—3 million acres, 14.4 million cubic yards of prime timber. That was the Chinese side of it—an area equal in size from Portland, Maine, to Detroit, Michigan, and from New York City south to Washington, DC. On the Russian side of the Amur, it was totally out of control and little was done about it. The Russians cared not a wit, believing that, by the time they are ready to harvest timber in this area, the forest will have regenerated. Thus there was no attempt to limit its movement. It burned for 45 days, and the Chinese estimate that the Russians lost ten times what they lost. But that is only a guesstimate. It could easily have been greater. The Soviets have yet to report the fire. Although the Chinese conflagration was picked up by orbiting satellite, it remains poorly known to the public and researchers alike (25).

The amount of CO_2 emitted to the atmosphere appears to have been colossal. Even so, from the Chinese, we also learn that between 1969 and 1986, another 5 million acres burned and that half of that burned twice.

And it was almost a year to the day that 500,000 acres of forest were lost to fire in Yellowstone National Park, in August 1988—minuscule by comparison to The Great Black Dragon Fire, but still contributing substantial amounts of CO_2. Of course, it will be years before the photosynthetic activity of these areas is regained. Thus loss by fire is a double shock to the atmosphere.

In a recent review and analysis of the atmospheric CO_2 burden, Keeling and co-workers suggest that since the 1970's burning and clearing of forests have become a greater contributor to the upsurge in CO_2 levels. Nevertheless, they would have expected still higher increases, well beyond the 350 ppm currently being recorded at Mauna Loa. To account for the "missing" CO_2, they postulated the operation of a countervailing "carbon dioxide fertilization effect." Keeling et al. conjecture that, when the air becomes saturated with CO_2, trees grow faster, thereby drawing additional CO_2 from the air as photosynthesis intensifies. But this remains to be ascertained. They also believe that the effects of deforestation are much greater than anyone has estimated, which means that increasing deforestation is much more dangerous to the climate than heretofore assumed (26). If the "fertilization effect" is correct, which ecologists have been quick to disavow, the remaining forests not only could be soaking up much more of the large additional quantities of the fossil fuel-derived CO_2, which could account for the small increases measured at Mauna Loa, but would tend to balance the global radiation budget. Thus, by saving the remaining forests and preventing further depredation, overheating within the greenhouse could be reduced substantially. If true, this conceptualization could be the most significant contribution to the dilemma of global warming. This idea would also lend great credence to the idea of massive reforestation.

Earlier, I asked whether an increase in global temperature to 68 or 69°F (20°C) would be desirable. Of course, that increase of only 5°C over the current temperature is the magnitude predicted for a warmer world. With the tracking of the trace gases over the past 30 years, it is reasonably clear that they are increasing substantially, given the record levels of use of fossil fuels, and that, if they continue to increase at current rates, their trapping of heat will ultimately destabilize the only humanly comfortable climate in the entire solar system. Indeed, for some scientists the rise has already begun. They maintain that, over the past 150 years, the earth has warmed 0.5°C. Others disagree. Nevertheless, an accumulating body of data suggests that if a temperature rise has not yet been observed, a clear signal will surely occur by the year 2040, 2050, or 2100, and that a series of untoward consequences will follow. What are we to believe? The first is that climate fluctuations are the rule rather than the exception. As shown in Figure 11 of Chapter 1, over the past 1 million years, the earth and its inhabitants have gone through at least ten shifts from cold and ice to cool and warm, and over the past 12,000–15,000 years, we've been in a humanly beneficial warm period. And, if nothing occurred to upset or alter these natural cycles of interglacial to glacial to current interglacial, the next cycle would be a cooling trend, as the earth enters a scheduled ice age. But something seems to be occurring that may not only derail the scheduled cooling and continue the warm period, but may shift it to an even warmer one.

How do we know this? The only way possible: climatologists have reconstructed the earth and its atmosphere and have injected CO_2 and other trace gases into their reconstructions to determine their effect on climate. These reconstructions are called models, and what models have been predicting have a number of scientists and political leaders worried and apprehensive. There is also little doubt that substantive information about how the earth–atmosphere system functions is being gained rapidly. But it is also evident that we are far from comprehensive knowledge of the workings of the natural carbon cycle and the mechanism that has maintained the global temperature stable at 59°F (15°C) all this time. Consequently, it is altogether fitting and proper to enquire if it is appropriate for humankind to place its trust in models . . . at this time. And it is also reasonable to question why credence is lent model projections of an anticipated temperature increase, as well as the time of its arrival, given past experience with models. Perhaps the next few questions should be: What are models, what can they tell us, and are they reliable? Or perhaps more to the point, what are the sources of their vulnerability?

REFERENCES

1. Michell, J. F. B. The "Greenhouse" Effect and Climate Change. Rev. Geophys. 27:115–139, 1989.
2. Fleagle, R. G., and Businger, J. A. An Introduction to Atmospheric Physics. Academic Press, Orlando, 1963.
3. Arrhenius, S. On the Influence of Carbonic Acid in the Air Upon the Temperature of the Ground. Philos. Mag. 41:237–276, 1896.
4. Keeling, C. D., Bacastrow, R. B., and Whorf, T. P. Measurements of the Concentration of Carbon Dioxide at Mauna Loa Observatory, Hawaii. In: William C. Clark (Ed.). Carbon Dioxide Review, Cambridge University Press, Cambridge, 1982, pp. 377–384.
5. Jones, P. D., Wigley, T. M. L., and Wright, P. B. Global Temperature Variations Between 1861 and 1984. Nature 322:430–434, 1986.
6. Jones, P. D., Wigley, T. M. L., Folland, C. K., Parker, D. E., Angell, J. K., Lebedeff, S., and Hansen, J. E. Evidence for Global Warming in the Past Decade. Nature 332:790, 1988.
7. Lorius, C., Jouzel, J., Ritz, C., Merlirat, L., Barkov, N. I., Korotkevich, Y. S., and Kotlyakov, V. M. A 150,000 year Climatic Record from Antarctic Ice. Nature 316:591–596, 1985.
8. Chappellaz, J., Barnola, J. M., Raynaud, D., Korotkevich, Y. S., and Lorius, C. Ice-Core Record of Atmospheric Methane over the Past 160,000 years. Nature 345:127–131, 1990.
9. Lorius, C., Jouzel, J., Raynaud, D., Hansen, J., and LeTreut, H. The Ice-Core Record: Climate Sensitivity and Future Greenhouse Warming. Nature 347:139–145, 1990.

10. Barnola, J. M., Raynaud, D., Korotkevich, Y. S., and Lorius, C. Vostok Ice Core Provides 160,000 year Record of Atmospheric CO_2. Nature 329:410–414, 1987.

11. Wigley, T. M. L., and Raper, S. C. B. Natural Variability of the Climate System and Detection of the Greenhouse Effect. Nature 344:324–327, 1990.

12. Study of Man's Impact on Climate (SMIC): Inadvertent Climate Modification. MIT Press, Cambridge, 1971.

13. Vaghjiani, G. L., and Ravishankara, A. R. New Measurement of the Rate Coefficient for the Reaction of OH with Methane. Nature 350:406–408, 1991.

14. Lovelock, J. Atmospheric Fluorine Compounds as Indicators of Air Movement. Nature 230:379, 1971.

15. Rowland, F. S., and Molina, M. J. Chlorofluoromethanes in the Environment. Rev. Geophys. Space Phys. 13:1–35, 1975.

16. Farman, J. C., Gardiner, B. G., and Shanklin, J. D. Large Losses of Total Ozone in Antarctica Reveal Seasonal CLO_x/NO_x Interaction. Nature 315:207–210, 1985.

17. Prather, M. J., and Watson, R. T. Stratospheric Ozone Depletion and Future Levels of Atmospheric Chlorine and Bromine. Nature 344:729–732, 1990.

18. Stevens, W. K. Ozone Layer Thinner, but Forces Are in Place for Slow Improvement. The New York Times, p. C4, April 9, 1991.

19. Lovelock, J. E., Maggs, R. J., and Wade, R. J. Halogenated Hydrocarbons in and over the Atlantic. Nature 241:194–196, 1973.

20. Weiss, R. F. The Temporal and Spatial Distribution of Tropospheric Nitrous Oxide. J. Geophys. Res. 86:7185–7195, 1981.

21. Coter, W. R. III, Levine, J. S., Winstead, E. L., and Stocks, B. J. New Estimates of Nitrous Oxide Emissions from Biomass Burning. Nature 349:689–691, 1991.

22. Thiemens, M. H., and Trogler, W. C. Nylon Production: An Unknown Source of Atmospheric Nitrous Oxide. Science 251:932–934, 1991.

23. Möller, F. On the Influence of Changes in the CO_2 Concentration in Air on the Radiation Balance of the Earth's Surface and on the Climate. J. Geophys. Res. 68:3877–3886, 1963.

24. Raval, A., and Ramanathan, V. Observational Determination of the Greenhouse Effect. Nature 342:758–761, 1989.

25. Salisbury, H. E. The Great Black Dragon Fire: A Chinese Inferno. Little, Brown, and Co., Boston, 1989.

26. Keeling, C. D., Bacastow, R. B., Carter, A. F., Piper, S. C., Whorf, T. P., Heiman, M., Mook, W. G., and Roeloffzen, H. A Three-Dimensional Model of Atmospheric CO_2 Transport Based on Observed Winds. 1. Analysis of Observational Data. Geophysical Monograph No. 55, pp. 165–236. In: Aspects of Climate Variability in the Pacific and Western Americas. American Geophysical Union, Washington, DC, 1989.

CHAPTER 3

MODELS AND MODEL PREDICTIONS

$\mathbf{H}$igh school and university students of the 1970s were nourished on the idea that a "population bomb" was about to go off. It was the biggest threat to life on earth, and if that didn't prove to be humanity's denouement, world famine surely would. Jay Forrester of MIT derived his doomsday predictions from a model he had constructed for The Club of Rome—a group of affluent international businesspeople who foresaw a bleak future for humankind. In 1972, the book *The Limits to Growth* by Donella and Dennis Meadows predicted dire threats facing the world and forecast the global depletion of such commercially important metals as gold, mercury, tin, zinc, copper, and lead, as well as the fossil fuels oil and gas, over the subsequent 20 years: gold by 1981, oil and natural gas by 1993, and in-between the others. In addition to a set of dismal predictions, their scenario, as with all such scenarios, was cast in a sense of urgency and high drama; of impending disaster if their warnings were not heeded now! And so they told us that "we have no doubt that if mankind is to embark on a new course, concerted international measures and joint long term planning will be necessary on a scale and scope without precedent." And they continued, "we are unanimously convinced that rapid, radical readdressment of the present unbalanced and dangerously deteriorating world situation is the primary task facing humanity." That was written in 1972. But it has a familiar ring to it. Unfortunately, apocalyptic predictions are a common feature on the American landscape. And the sky is always about ready to fall in. So it was with the Meadows—urgency was required if Armageddon was to be avoided. "We can talk about where to start," they admonished us, "only when the message of *The Limits to Growth*, and its sense of extreme urgency, are accepted by a large body of scientific, political, and popular opinion in many countries. The transition in any case is likely to be painful, and it will make extreme demands on human ingenuity and determination. As we have mentioned, only the conviction that there is no other avenue to survival can liberate the moral, intellectual, and creative forces required to initiate this unprecedented human undertaking." And finally, "the last thought we wish to offer is that man must explore himself—his goals and values—as much as the world he seeks to change. The dedication to both tasks must be unending. The crux of the matter is not only whether the human species will survive, but even more whether it can survive without falling into a state of worthless existence." This is what students and the public were asked to believe. Those who write this type of purple prose ought to be condemned to reread it at least once each year. This scenario had followed the publication in 1968 of *The Population Bomb* by Paul Ehrlich, which predicted worldwide famine and death by the early 1980s as a consequence of overpopulation and declining food supplies. Now, 20 years later, neither famine, depleted resources, nor overpopulation have encumbered the world. In fact, many countries must deal with the twin issues of oversupply as well as a declining population base, and with it a bleak outlook for fulfilling their needs for an adequate labor force. "Bombs" of neither type have, or are about to, go off.

The problem with doomsayers and catastrophists, flitting as they do from crisis to crisis, is that they are rarely required to confront their predictions. Thus Julian L. Simon may have provided a model for prophet testing. In 1980, Simon, an economist thoroughly displeased with Ehrlich, a biologist and champion of doom, bet Ehrlich $1000 that his predictions of a population explosion with its attendant ramifications were wrong. They bet on the future price of five metals—copper, nickel, tin, tungsten, and chromium—in quantities that each cost $200 on the current market; and each agreed to sell the other the same quantity of metals 10 years later based on 1980 prices. If 1990 prices were higher than $1000, Simon would pay Ehrlich the difference. If the 1990 prices were lower, Ehrlich would have to ante up. As Simon saw it, prices would fall as technological advances and good management would unearth new sources and retrieve abandoned ones. According to Ehrlich, as population increased resources must become scarce and prices rise. Actually, their bet was over their views of the planet—progress versus doom. In 1990 the bet came due. Although over the decade 800 million people were added to the world's population, the prices of each of the metals declined—substantially. Ehrlich had to pay Simon almost $600. Over the 10 years, finds of new mineral deposits and new processes of extraction and refining brought more and cheaper minerals to the marketplace. In addition, other materials replaced the traditional ones—copper, for example, is being replaced in telephone lines by fiber optics, and ceramics are replacing tungsten. It was ever thus. But for some this ancient lesson will never be learned. It is easier to predict catastrophes. Simon has suggested another bet, this time for $20,000. Ehrlich has refused the offer.

The students of the 1970s may recall their dread of lives in turmoil or cut short by two of the apocalyptic horsemen. Now, at 30 and 40, they face a new threat; computer-generated models warn of wholesale ecologic dislocations as a result of rising global temperature. Once again jeopardy stalks their lives. However, given these recent false prophecies, a fair degree of skepticism about predictions and consequences of a warmer earth would be understandable. After all, each of the earlier dire forecasts had, as does global warming, the imprimatur of a segment of the scientific community. Given public response, however, to the news of trouble in the "greenhouse," it is apparent that some remain highly susceptible to the predictions, while others are clearly immune; but for the greatest number their responses are yet to become manifest. That is, sensitivity to models remains questionable or unknown, which raises the further question of what can or should be expected of models, and in the special case of climate models is more or less to be expected.

At one time or another we've all constructed models—planes, trains, ships, animals, birds, houses, cars—our attempts to portray or simulate reality in a manageable and inexpensive way. Indeed, marine engineers can build a model of a ship, and aviation engineers can build a working model of a plane, and both can be tested—one in a tank of water, the other in a wind tunnel—wherein a variety of forces can be applied to each. Waves can be generated and sent crashing against a newly designed hull, and winds of hundreds of miles per hour be hurled at wings. Will the ship list and capsize, will the plane lose speed and altitude? Models can tell us a great many things quickly, safely, and reliably.

But models are not limited to objects that can be carved or glued together; models can also be concepts that can be verbalized or expressed mathematically. The objective of each is to provide a framework within which the results of alternative assumptions can be identified, but they are not substitutes for choosing the assumptions, relationships, and data needed to describe the system being modeled. They are constructed to reflect the real world where they can and incorporate assumptions where they cannot, in a consistent framework. The mathematical model goes beyond the conceptual by placing numbers on variables and determining, in some instances, rates of change over time; it becomes quantitative rather than simply descriptive, and as the model itself increases in complexity it can provide information on the rates of change of several mutually dependent variables concurrently. Thus it can be as simple as a single equation relating a dependent variable to an independent one: the numbers of cigarettes smoked and the development of lung cancer. It can be moderately more demanding in attempting to predict the traffic flow out of Manhattan via all its roads, bridges, and tunnels in response to an explosion at a nearby nuclear power plant, or more demanding yet by attempting to forecast the weather over the southeastern states 30 days from now. Of course, problems are to be expected as the degree of complexity increases, especially where natural forces are involved. Nevertheless, over the past 500 years many of nature's secrets have been discovered, so that humankind can now overcome the forces of gravity and place a person on the moon—which requires knowledge of planetary motion and the physical laws of momentum and gravitation. These have been reduced to mathematical equations, which have predicted that if an object, a spacecraft, were hurled upward with a specific velocity, it would exceed the earth's gravitational pull and escape into space. And having also predicted the location of the moon at a specific time, the spacecraft moving at a known speed would arrive at a point in space at the same time as the moon. And so it was that 102 hours after takeoff, Neil Armstrong and Edwin Aldrin, Jr. arrived at the appointed spot in their *Apollo 11* spacecraft and descended onto the moon at 10:56 p.m. on July 20, 1969. It works.

$E = mc^2$ is a model. And it works. In its consummate elegance it expresses the formidable relationship between energy and matter. In the Special Theory of Relativity, $E = mc^2$ identifies energy with mass and in so doing does not violate the Laws of Conservation of Energy, the physical principles describing the transfer of energy. Thus the conversion of a small amount of mass can yield a very large amount of energy. Since c^2 is a very large number and even if m is small, E will always be large as we have seen when the atomic bomb was detonated at White Sands, New Mexico.

With the available supercomputers, a modeler can simulate the thrust of a rocket engine, calculate the heat and pressure on a missile's warhead as it enters the atmosphere, and predict all the forces affecting it from launch to impact without actually firing one over the earth. And having programmed a computer with the requisite equations, it can simulate the implosive shock wave that detonates nuclear warheads, calculate the multiplication of neutrons in a chain reaction, and model the complete process of nuclear fusion of a hydrogen bomb. Indeed models can and do work—with good reason. Calculus is the reason.

Calculus is the mathematics of continuous change, for which we owe Isaac Newton an everlasting debt. The idea that change or motion can be represented by mathematical equations is the essence of computer modeling of global climate. Herein is concern for an ever changing fluidlike atmosphere with a broad range of interactive elements. If the factors or climate variables can be represented by an equation, then calculus can deal with them via differentiation and intergration; differentiation computes the rate at which one variable changes in relation to another at any instant, while integration takes an equation in terms of rate of change and converts it into an equation in terms of the variables that do the changing. The logic of calculus can be applied to any situation or condition that can be reduced to an appropriate mathematical statement or set of equations. It works wondrously. Evidence of continually varying change, if needed, surrounds us: automobiles moving along at frequently changing speeds, accelerating and decelerating; birds in flight, diving and climbing, changing direction; a baseball player preparing to steal a base; the wind on a balmy day. Everything about us changes constantly. In the words of the ancient Greeks, "all things flow."

The atmosphere and oceans, elements of climate, are in constant motion, ever changing; and if the foregoing is correct, the elements of climate should be reducible to mathematical statements and their solutions, providing descriptions of climate over time. It makes sense. Figure 1 suggests the type of linkages that together may produce worldwide climate. Before moving on, it would be worth a moment to reflect on the ideas "suggest" and "may." Does certitude have a place here? In more graphic form, Figure 2 displays a rendering of the complex web of activities affecting sources and sinks of the radiatively active trace gases. It goes without saying that for meaningful results, these too must not only be known, but known with some precision. From the two figures alone it would not be incorrect to consider that the scope of the undertaking borders on the overwhelming. Quite obviously, these interlocking systems are far too complex to be described in one or two sets of equations. Here then is the place where numeric models find applicability.

Calculus has worked for mankind for over 250 years. It needs no defense. But it is worth remarking that if, say, a model's description of drag forces on a plane's wing or the stability of a car-ferry running in high seas proves less than satisfactory during trial runs, reprogramming is no problem. But climate cannot be studied under controlled conditions in the field or laboratory. It will not hold still for examination. Nevertheless, questions must be asked, and one of the more fundamental is this: Is the nature of climate, the process, as Figures 1 and 2 attempt to set forth, sufficiently well understood to reduce to an appropriate set of equations? That's a key. How much comprehension is necessary? Can we obtain sufficient data for our needs from less than full knowledge? Considering that the first climate model was developed in 1963, the science is not yet 30 years old. And being under 30, why, one wonders, is so much faith vested in it? The answer is not difficult. How, other than by simulation, can this vast natural phenomenon be studied? There is no other way. In a word, it's all we've got. Archimedes understood this conundrum. "Give me a place to stand," he quipped, "and a staff long enough, and I'll move the world." But is availability sufficient?

Models can also be wrong—dead wrong—as was recently shown in the Persian Gulf War. Before the massive raids on Iraq began, the Tactical Air Command Center in Riyadh took many steps to protect the raiders—pilots and planes. Special Force commandos were sent into Iraq and Kuwait to locate and identify early warning and fire control systems. Attack bombers followed up and destroyed them. Stealth bombers hit heavily defended targets near Baghdad. These precautions worked beautifully. After the raids began, all the planes returned safely. Not one was lost. Yet in simulations before the attack, Air Force computers had projected that as many as 150 planes would be lost in the first night. As with climate projections, models may well be overly sensitive.

Because relationships can be expressed in the language of mathematics does not mean they are without fault or error. Predicting a population explosion, and with it worldwide famine, is an excellent example of a problem involving a daunting complex of constantly changing interre-

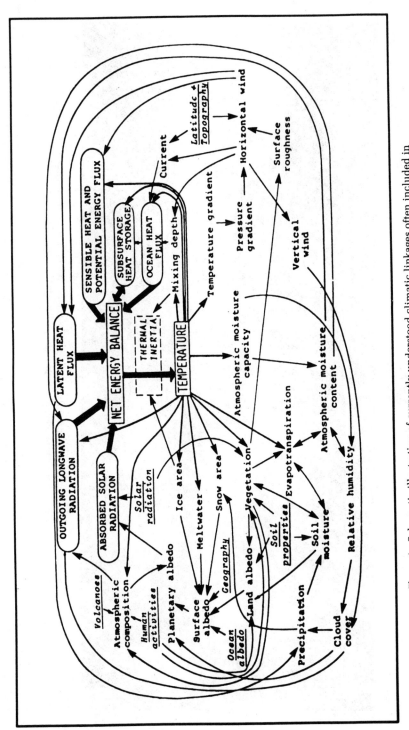

Figure 1. Schematic illustration of currently understood climatic linkages often included in numerical models. (From Projecting the Climatic Effects of Increasing Carbon Dioxide, U.S. DOE, ER-0237, Dec. 1985.)

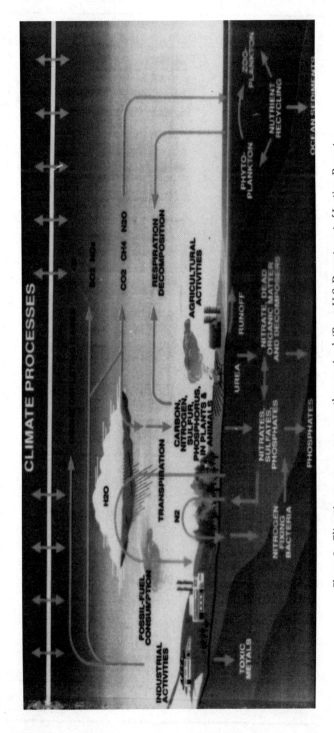

Figure 2. Climatic processes as currently perceived. (From U.S. Department of Justice, Report of the Task Force on the Comprehensive Approach to Climatic Change.) (See color section.)

lationships, which were expressed mathematically, modeled, and depicted as a set of predictions—which were wrong. Thus it is important to understand that models do have limitations. They cannot incorporate numerous factors that are yet unidentified or poorly understood and be expected to yield faithful forecasts. Furthermore, efforts to predict next year's climate have proved fruitless, and currently there is a total inability to forecast unusual changes. For example, from January 25 through February 1990, hurricane force winds often exceeding 100 miles per hour ripped through western Europe, unmercifully battering cities from Scotland to France, taking 140 lives and exacting damage to property upward of a billion dollars. A similar storm assaulted the continent from Sweden to Spain in the fall of 1989, and the most severe storm of the 20th century struck an unprepared Europe early in January 1976. None of these were forecast. And as you well recall, econometric models fared no better. Black Friday, October 19, 1988, will long be remembered on Wall Street and Main Street as the day stock prices dropped some 500 points with a loss to investors of half a trillion dollars. Not only was this not forecast, but for the entire day events could not be estimated or managed. As David Rind and Arthur and Cynthia Rosenzweig have suggested, "the inability to foresee these extreme anomalies has properly raised doubts about how well experts in both fields understand their subject, especially when dealing with situations outside the norm" (1).

We know our local weather. We know too that our climate changes with the seasons and we live by these changes; but we also know that within each season there will be unexpected excessive snowfalls, rainfall, and heat, yet we don't know when those excessive periods will occur and we cannot predict any of them for the next week, month, year, or decade. We simply do not know enough of the workings of the process or system we call climate. We don't know if the process contains elements of randomness, or if the system is totally and tightly coupled. But this grand lack of ability to forecast the weather does not mean that predictions of a warmer world are necessarily wrong or baseless. At this juncture models provide the only way in which climate can be studied under even a semblance of controlled conditions.

If uninterrupted, the present warm interglacial period should begin to cool over the next 2000 years as the earth moves into the next natural glacial period as the Milankovitch cycles suggest; and as it cools there should be a decline in atmospheric levels of CO_2. If, however, large additions of CO_2 from burning fossil fuels and deforestation interdict the process, an extended superinterglacial period has been predicted to occur, lasting another thousand years, but with surface temperatures higher than the earth has previously known. The problem for humankind is that there is no precedent, no experience, no guideline for the type of climate change and natural dislocations that could ensue. The best that can be done is to attempt to predict the future, which is always very dicey—especially so when the climate system is so vast and so complex, and our knowledge of it sparse and therefore uncertain. Yet the only applicable method for projecting future climate is the construction of a climate model based on physical principles known to govern the system. As if that were not sufficiently unsettling, the predictions and projections emanating from the models have little or no way of being objectively verified—except by other models. It is within this highly uncertain realm that climatologists seek to extract certainty, or something approaching it.

The object of climate modeling is the application of fundamental physical principles such as the laws of mass, momentum, and energy to the spectrum of climate components currently known in some measure, such as the concentration of gases making up the atmosphere, ice coverage over the earth, and interrelationships between the biosphere and atmosphere and the oceans. The model is essentially then a set of physicochemical relationships expressed as mathematical equations, which attempt to represent the interactions between components of the natural system, and which when given appropriate commands can be solved obtaining simulations of climate under a variety of "experimental" conditions. But the models are also expressions of assumptions about the real world, again in the form of equations. The key or operational idea, which must not be glossed over or taken lightly, is assumptions. The questions that all modelers face is what to do when firm real world data are either unavailable or lacking. Two choices present themselves: omit or assume. Guess. Fill in the void. Both nature and modelers abhor voids. Fill in the void with something reasonable.

If climate is not a random event, and little exists to suggest that it is, then forecasting the climate should be possible, if we know the plethora of variables that constitute the process. That is, given a set of conditions, it should be possible to predict climate. Unfortunately, at this stage of its development climatology, the science of climate, does not possess that high degree of knowledge. Thus the need for models. And given the system's astonishing complexity, computers, more often than not supercomputers, are required to solve the equations.

But it does not necessarily follow that a direct relationship exists between the validity of a model (compared to observed reality) and its complexity. Neither do computer size and validity necessarily go together. The real question or test of a model is the closeness with which it approximates reality. Of course, it is the large, abrupt change in temperature that will be the greatest problem for humanity. Small changes over long periods can be managed and adjusted to: not so with major shifts over a short time frame. At this point, then, we are in the hands of the modelers, for there is no other forecasting mechanism on which to rely. We must hope that within the next 10 years uncertainty gives way and we can place our trust in their projections.

The question at issue is what will a doubling of trace gases do to our normally stable climate? This relates directly to the sensitivity of the system and is the type of question appropriate for models. Given the immense complexity of the climate system, many of the models view it as being divisible into three boxes—an atmosphere, a biosphere, and an ocean. The problem the model wants to solve is what the relationships are between the boxes, what controls the effects between the boxes, and how they are affected unnaturally, say, by large injections of radiative gases or volcanic ash into the atmosphere. It should be evident that for the model to resolve these questions, each of the boxes must be well defined and understood. It is here, all too often, that lack of knowledge is replaced by uncertainty and assumptions. After establishing the blocks, boundary conditions are imposed. These are the existing environmental conditions under which the model will run.

In fact, the earth is divided into much smaller boxes for ease of management. Arguably the most sophisticated of current generation supercomputers is the CRAY XMP. General circulation models run on these machines are limited in their resolving power. Consequently, global climate rather than regional or local weather is all that can be managed at this time, and that with inherent uncertainty. Calculations are based on widely spaced points on a three-dimensional grid overlaying the surface of the earth. Figure 3 shows the grid as well as an exploded view of a single box with its nine-level elevation, with grid lines 400 kilometers by 400 kilometers. For relative ease of managing the data, the surface has been divided into 1920 such rectangular "boxes." With nine levels to a height of 1–5 kilometers (1920×9), 17,280 boxes or climate units interact to give a "picture" of world climate. The models must then solve simultaneous equations (a set of equations that are all satisfied by the same values as the variables, and in which the number of variables is equal to the number of equations) for the climate variables in three dimensions. It is at the intersection of the grid points that the complexity of weather and horizontal variation is reduced to a single value for temperature, cloudiness, wind velocity, precipitation, sunlight, soil moisture, and other factors each modeler chooses. Thus from grid point to grid point, there are of necessity huge gaps, and within each box there is only one value or number representing the weather conditions within those 160,000 square miles—an area the size of Colorado. And among modelers physical processes are represented differently. Different schemes are used for computing the height, cover, and optical properties of clouds, the hydrologic cycle, sea ice, surface albedo (reflectivity), seasonal cycles, and, possibly most important, the treatment of the oceans. Accordingly, differences in model predictions must be seen as the rule rather than the exception. And until a computer can handle 50–100 times as many discrete boxes, local weather patterns will not be resolvable. In large measure, because clouds play so great a role

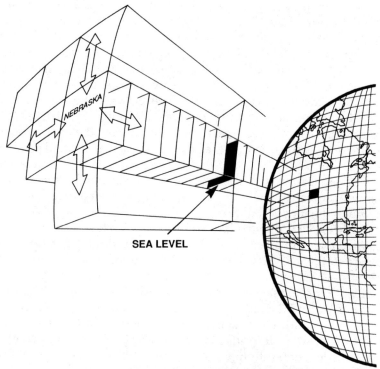

Figure 3. Modelers often divide the earth's surface and the atmosphere into "boxes" or levels to more effectively manage their data. The exploded view here suggests the degree of complexity of the interacting variables.

in local weather, and because they exist within a few kilometers, we cannot expect models to give more than they currently do, until such time as they can resolve kilometers rather than hundreds of kilometers.

As Figure 1 suggests, the interconnections are awesome, and their constantly changing nature assures that the response of the system will be staggeringly complex. To deal with this complexity and help fill the void in our understanding of how planetary climate responds to a change in one, two, or more variables, atmospheric physicists, oceanographers, biologists, and geologists have developed a number of increasingly complex models, which attempt to simulate or reproduce the real world by representing mathematically what is known about each of the component parts of the system. All models can be subsumed under one of two major types: hydrodynamic or thermodynamic. The difference between them depends on whether the conversion between kinetic energy and total potential energy is or is not included. Thermodynamic models include energy balance and radiative–convective models, while the hydrodynamic models encompass atmospheric general circulation models and statistical–

dynamical models, relatively simplified models constructed on versions of the equations for conservation of energy and mass and averaged over longitude, with the effects of swirling air masses—eddies—over large areas estimated along meridians. The simplest (of the highly complex) models are the one-dimensional radiative–convective models, which seek to describe the vertical structure of atmospheric temperature. Energy balance models attempt to calculate surface temperature in terms of a balance between incoming solar and outgoing infrared radiation. And the most sophisticated are the three-dimensional atmospheric general circulation models (AGCM) in which the atmospheric circulation is simulated over an extended period by means of time-dependent sets of equations that describe the changing thermodynamic state of the atmosphere with latitude, longitude, and altitude. Essentially, a zero-dimensional model might attempt to predict the change in a single climatic element without concern for a time frame—for example, the determination of an equilibrium temperature given the known insolation (incoming solar radiation). A one-dimensional model would add the extra dimension of, say, altitude. If to this temperature determination were added the temperatures at a range of locations around the world, a second dimension would have been added. And finally, the most complex or sophisticated model, the three dimensional, adds yet another stratum by incorporating temperature readings at a range of altitudes along with a range of locations—both north/south as well as east/west.

With these numerical equations the models generate a sequence of maps showing variations in worldwide climate consistent with a specific set of incorporated boundary conditions. Figure 4 is a set of computer simulations of greenhouse warming in July of six years—1986, 1987, 1990, 2000, 2015, 2029. Red is hotter than "normal," the 1950–1980 average, and blue is colder than normal. What is seen is that by the 1990s the area with above-normal temperature begins to exceed the area with below-normal temperature. This is what models can do. Exact patterns for specific areas cannot be generated. General ideas can be generated.

In this regard, Manabe and Wetherald of the Geophysical Fluid Dynamics Laboratory/NOAA, in what stands today as a pioneering and classic climate modeling study, found that clouds tend both to cool the earth by reflecting sun back into space and also to warm it by intensifying the greenhouse effect. By varying the amount of cloud cover at a number of altitudes, their model responded by indicating that the reflecting effect, cooling, was greater than the warming. Consequently, clouds appeared to be a negative feedback mechanism tending to keep the earth cool (2). Perhaps the most significant advance in modeling of temperature change, with concurrent increases in CO_2 emissions, was the development of the three-dimensional climate model, again by Syukuro Manabe and Richard Wetherald (3). Although quite limited by comparison with current GCMs, it provided for the first time an indication of how increased CO_2 might

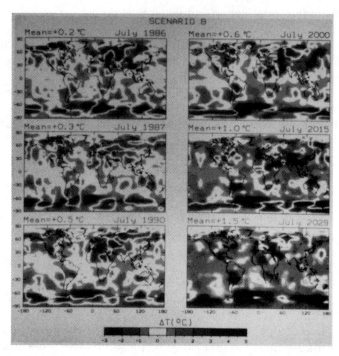

Figure 4. Simulated July temperature anomalies for the 6 years indicated. The simulation is based on a specific set of boundary conditions. (Courtesy of James E. Hansen, Goddard Institute for Space Studies, New York.) (See color section.)

affect atmospheric temperature, rather than being limited to describing surface temperature. In addition, again for the first time, they provided an indication of the manner in which the earth's water balance might be disrupted.

In August 1988, Hansen and co-workers published the results of a simulation using the Goddard Institute for Space Studies Model II GCM, in which CO_2 was instantaneously doubled. They produced three potential scenarios: scenario A assumed continual exponential growth of trace gases; scenario B assumed reduced linear growth; and scenario C assumed a rapid curtailment, such that after the year 2000 the net climatic forcing by CO_2 ceases. Included too were potential contributions of other trace gases. They concluded that scenario B was the most likely or plausible outcome, and they noted that the decade of the 1990s would be clearly identifiable as the period when greenhouse worry become apparent. But they also added a caveat. They called for more research to obtain the type of measurements needed to improve current knowledge of climate forcing (4).

The structure of one form of an AGCM model with its various components is shown in Figure 5. It consists of five elements: the equations of state, of motion, thermodynamics, radiative transfer, and water balance. Thus, in addition to statements (equations) about the components of the climate system, the model is underpinned by the laws of conservation of energy, momentum, and mass, equations dealing with rigorously established natural phenomena. And to demonstrate the ability of the model to simulate the climate, the global distribution of the rate of precipitation projected by the model is compared with figures for observed precipitation. Models appear to reproduce this reasonably well.

Obviously, the models require detailed knowledge of atmospheric physics and chemistry, biology, hydrology, mathematics, and other disciplines. Consequently, for models to be accurate and dependable instruments, interdisciplinary teams are needed for their development and interpretation, or at least these teams must communicate effectively. The above notwithstanding, uncertainty is inherent in all models, and uncertainty means models cannot yet replicate the real world; but they may come close. The questions are how close and is close adequate. These are two of the nagging questions and will remain so until there are suitable means for verifying model predictions. There is yet another element that comes into play; this is the sensitivity of the simulated system. Sensitivity refers not to

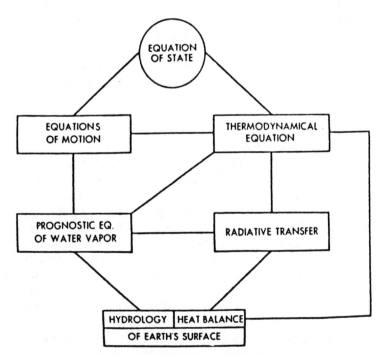

Figure 5. The box diagram shows the basic components of an atmospheric model along with their perceived interactions. (Courtesy of S. Manabe, NOAA/Geophysical Fluid Dynamics Laboratory, Princeton, NJ.)

whether the artificial system can wholly articulate the real world, but rather to what degree the simulated system's responses match the real system. Models attempt this by trying to reproduce this year's climate, or at least the fluctuation between winter and summer, which is by far the greatest. How well they can match that large difference is seen as a reasonable test of their sensitivity. In addition, as the years pass and climate changes, or doesn't, modelers will be able to fine tune their models and further verify their results.

Uncertainty was no more strikingly exemplified than in the vigorous rebuttal to the certainty that global warming was upon us. The summer of 1988 was hot. The excessive heat was unabating and drought was upon the Great Plains. In April, Hansen and Lebedeff of the Goddard Institute for Space Studies (GISS) had published their data on global mean surface temperature changes over the years 1880–1987 (5). The journal in which it appeared was sufficiently esoteric to assure that only like-minded scientists would see it. But that wasn't the end of it. In May, Hansen testified before a Senate committee and in strident tones maintained that the greenhouse warming was at hand. It became a media event, especially as the drought continued. The GISS model, one of the most sophisticated of the GCMs, had indicated an increased frequency of droughts and hotter than normal summers. It could not have appeared at a more opportune time. Figure 6 describes the trend in temperature change. The data points are based on records of land-based temperature stations and mean to show that surface air temperatures in the 1980s were the warmest in the history of instrumental records, and that the rate of warming between the 1970s and the late 1980s was greater than that which occurred between 1880 and 1940, an earlier period of warming. Both the annual mean temperatures and the 5-year average clearly show this upward trend—of 0.6 or 0.7°C. Allowing for urban heat island effects of some 0.1–0.2°C, they estimated that the 1987 global surface temperature was 0.63 ± 0.2°C warmer than in the 1880s. Hansen maintained he was 99% confident the earth was indeed getting warmer, and that the warming was probably due to the greenhouse effect. These remarks brought a storm of protest from the scientific community, who believed that uncertain data were being given greater certainty than warranted. Figure 7 traces surface temperature from 1860 to 1990, at which point predictions suggest there may be three ways to go: upward for an additional 0.8°C/decade mean temperature increase, if nothing is done to curtail greenhouse gases, upward to a 0.3°C/decade if some restrictions are imposed, or up no more than 0.6°C/decade by the year 2040 if emissions are curtailed. The essential point is that for some modelers there is no question that warming will occur. The issue is only of degree.

More recently, Hansen could quite reasonably invoke the work of Kuo, Lindberg, and Thompson in support of his warming argument. "Kuo et al.," he said, "concluded that the probability that the data, at face value, show a true warming trend, rather than a statistical fluke, is 99.99%. But whether it is 99.99% or 99% or 95% is not really the point. The point is that

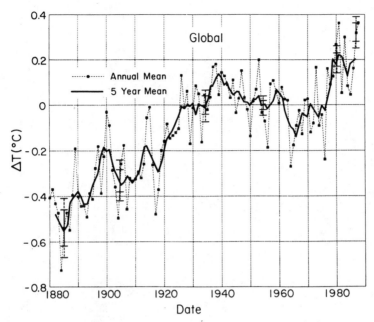

Figure 6. Meteorological station temperature data between 1880 and 1985. The data suggest that the 20th century is warmer than the last half of the 19th century. The solid line represents 5-year averages. (Courtesy of James E. Hansen, Goddard Institute for Space Studies, New York.)

the world is almost certainly getting warmer. That is a solid scientific conclusion." Perhaps. But Kuo and her colleagues of the Mathematical Sciences Research Center, AT&T Bell Laboratories, had in fact this to say: "Present numerical models are crude, and comparisons between the time series representing the real data and predictions of the atmospheric models are difficult to interpret. Because the available data are short-time series, conventional statistical methods are unreliable, and detection of the greenhouse effect remains controversial" (6).

This is a critical consideration, as the threat of an increase in global mean temperature is probably the gravest threat faced by mankind since hunter/gatherers became farmers and established communities some 10,000 years ago. To deal with a threat of so great a magnitude, we ought in fact to know the magnitude of the threat with some certainty. Consequently, the reliance on models becomes of singular importance because they are all we have to rely on. The question of whether or not the empirical data—the observational data, in fact—show the warming trend becomes central to any discussion. This is the issue for Kuo and her coworkers. Thus they maintain that their analysis of the Hansen–Lebedeff data shows "that from 1880 to 1987 the average global temperature increased by 0.0055 ± 0.00096°C/year, and the probability that this slope is positive exceeds 99.99%." They continue with the statement that "the

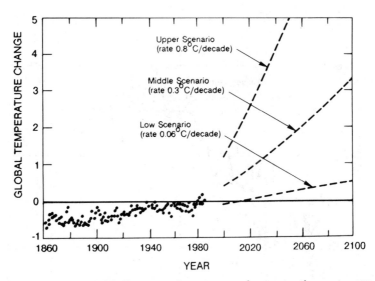

Figure 7. Three scenarios of globally averaged temperature change over the coming 120 years are presented. Each represents a level of carbon dioxide emission and its relationship to an anticipated temperature level (42).

coherence results presented here provide significant evidence that the average global temperature and CO_2 concentration from 1958 to 1988 are linearly related at many frequencies." Having established this relationship, they then raise a significant caveat and go on to warn readers that "caution must be exercised in interpreting this result as suggesting that the variations in atmospheric CO_2 are causing the changes in global temperature, even though there are plausible physical mechanisms linking the two series. Apparent correlations that are used to postulate causality can sometimes be misleading. . . . In addition, we should be particularly cautious in interpreting the coherence when the series analyzed are as short as these: climatic and solar variations often are of longer duration than these records." It is incumbent upon the reader to recall that the Hansen–Lebedeff temperature data are all from land-based stations, which means that the 72% of the global surface covered by water is not included. And furthermore, time series in terms of cause and effect have built-in weaknesses. The fact that the Hansen temperature series and the Keeling CO_2 series have positive slopes may or may not be cause–effect evidence of a warming trend precipitated by increased CO_2. Although they could very well represent a cause–effect relationship, causality does require that one precede the other: in this case it would be nice if CO_2 increases preceded the temperature increases. If a degree of skepticism is extant, then one would be constrained to note that many variables can be shown to move together in time but are unrelated causally. For example, the incidence of lung cancer has risen steadily since World War I, when the Red Cross supplied the newly mass produced cigarettes to the soldiers in France. From 1918 to 1980 the trend lines for asphalt use, creosoted telephone poles, and the importation of bananas have all risen along with the number of cigarettes smoked, but none of these has been seriously considered as a risk factor in the causation of lung cancer. By the same token, cigarette smoking by men rose steadily, followed—after a latency period of 20 years—by an increasing incidence of lung cancer. This could be viewed as weak circumstantial evidence of a cigarette–lung cancer association. When the trend line for women and lung cancer began to rise as it did for men, after a lag of 40 years, the evidence for a cause–effect relationship between smoking and lung cancer was further strengthened. In addition, when male physicians as a group stopped smoking and their lung cancer death rates declined, that added yet greater strength to the association. However, the association between increasing surface temperature and increasing levels of atmospheric CO_2 lacks any strengthening relationships. Of course, it is a relationship for which it is exceedingly difficult to test or to develop supporting data. Thus we are at an intellectual impasse: it may well be a fact; but it also may not be. The path to follow for decisionmaking may thus be debated more by personalities than by evidence—leaving a good deal to be desired.

Major uncertainties in model calculations arise from the severe limitations of existing knowledge of what makes climate—the mechanism—as well as the uncertainty in representing the interactions of the multifactorial process in a model. Figure 1 serves to accent the complexity. It should be given more than a cursory inspection, as it diagrams a welter of perceived interactions. Obviously, some have received greater attention over time than others; thus more realistic data can be called upon. The arrows suggest the direction of interplay, but whether one component plays a greater role is at this time only an assumption. It should also be recalled that we must deal with collections of worldwide data. That is, the interactions depicted in the figure must be studied and known not only over Nebraska and California but over New Delhi and Tokyo as well, and that immediately suggests that much has never been collected or has been collected only recently. But most evident are the extraordinary linkages and feedbacks, in which a change in one could amplify or moderate another or others, producing either positive or negative feedbacks. Does the figure represent the totality of interactions that together produce global climate? It would be useful, and comforting, to know that.

THERMOMETERS AND TEMPERATURE

Data over the long term are essential if trends are to be discerned. Unfortunately, representative historical data have either not been collected, are in unusable condition, or are unavailable. In place of worldwide data, modelers have used northern hemisphere land-based records to simulate the world. Filling in voids by making "reasonable" assumptions, or "parameterizations" as they are called, can introduce error and thereby uncertainty. Land-based data may simply not fill in for the oceans, which cover 72% of the earth's surface and which have far greater heat capacity than the land. In addition, since the 1860s the means of obtaining sea-surface temperature have shifted from "the bucket method" to thermometers located in cooling water intakes, which yield slightly higher readings. In fact, these slightly warmer recordings average about 0.5°C higher —just about the amount of temperature shift estimated by models to have occurred over the past 100 plus years. Furthermore, over the same period many of the formerly isolated worldwide temperature stations have become surrounded by encroaching urban growth, which itself induces a "heat island effect" sufficient to confound actual surface temperatures from the additional heat contributed by the energy sources of urban areas. Additional confounding occurs as a result of a program of replacing instrument shelters at airports around the country with remote reading thermometers, which also involved changes in location and elevation. It has been found, for example, from overlapped data from both the newer and older instruments that the changeover was responsible for changes in monthly mean temperatures "by amounts exceeding 2°F (1.11°C) in many cases and 3°F (1.67°C) in a few cases" (7). But his cautionary note was striking "The dozens of extreme records, that recently have been broken," he wrote, "were not the result of a sudden shift in climate, but of a change in exposure for a large part of the network." How widely this finding was disseminated, and whether it served to revise data bases, are yet open questions.

Another questionable practice involving temperature station data is its time of reading or collection. Although early in the 20th century the U.S. Weather Bureau introduced a standard daily mean temperature as the average of the maximum and minimum daily termperatures, and although this was widely adopted, differences persisted. For example, whereas weather stations defined the day as the 24-hour period immediately preceding the observation, many simply made their observation at their convenience. Few, of course, chose midnight; more suitable hours were between 7 and 8 a.m., the late afternoon, and sunset—which itself varies widely. These differences in daily observation affect both monthly and annual mean temperatures; some will yield monthly means that are too cold and some that will be too warm. Thus the true mean of any station may well be in doubt. Again, with half a degree Celsius as the calculated increase over the past 100 years, these small errors take on a larger significance. Imagination may be a substitute for voids in data, and that is perfectly reasonable in experimentation; but imagination is not truth. The distinction must not be confounded.

An example of misleading temperature data was recently publicized by the National Weather Service*. They are checking the possibility that record-breaking heat in Tucson, Arizona in the summers of 1987, 1988, and 1989 was in fact due to new but faulty thermometers, which on 15 separate occasions over the 3 years registered temperatures above the city's previous high of 111°F. In the summer of 1989, an all-time high of 117°F was recorded. In other cities, these new thermometers had been recording temperatures 3 degrees higher than other instruments. The new thermometers are electronic sensors called HO83, which the Weather Service is counting on to replace the traditional mercury and glass units. These sensors record instantaneous changes in temperature and keep track of their own observations, which obviate the need for hourly or daily readings by a person using the standard mercury thermometer. The problem of the sensor may be faulty design in that its fan is not large enough and allows absorption of too much solar radiation. This suggests that the air temperature being measured has been artificially heated. If we are to determine if the earth is actually getting warmer, we must be certain our recording instruments are yielding accurate and reliable data. It is noteworthy that mercury thermometers have an error range of three-tenths of a degree.

For some, it may be irrelevant if the temperature is recorded as 111°F or 115°F; either one is too hot. Unfortunately, it does matter, especially when mean global temperature is in question. Erroneous local recordings will play havoc with estimation of an average worldwide temperature— especially when attempting to determine if over the past 100 years global temperature has in fact increased 0.3–0.5°C. The fact is that climate models have predicted that warming, should it occur, will not be evenly distributed. The highest increases are predicted for the polar and subpolar regions, where temperatures are predicted to rise as much as 8–9°C. Thus it would be of the utmost importance to obtain ongoing temperature data in these regions. Hence the report on the current ocean temperature in the

---

*As early as 1797, Thomas Jefferson envisioned a nationwide network of volunteer weather observers. From 1776 to 1816 he had set up a system of observers in every county in Virginia. By 1800 there were volunteer stations in six states: Massachusetts, Pennsylvania, Connecticut, New York, North Carolina, and Virginia. But it was not until 1953 when Dr. Helmut Lansberg of the Weather Bureau developed a national east/west, north/south grid with stations every 25 miles that country-wide coverage became possible. Currently observers at some 10,000 stations record temperature and rainfall data. Many have made daily measurements for 30, 40, and 50 years.

The longest continual record of observation by an individual is held by Edward G. Stoll, who has taken readings for 76 years in Arapahoe, Nebraska. But in July, 1991, the National Weather Service honored John W. Maddox of Rome, Georgia, for operating a weather station began by his family 136 years ago that has operated continuously ever since.

Over this time, however, none of the weather services has ever developed a standard procedure for reading thermometers. Consequently quality control has always been in the eye of the beholder.

area of the North Pole by a team of German, Canadian, and American scientists becomes uniquely pertinent. They collected data aboard the Soviet ice-breaker, *Rossiya*, on its way from Murmansk to the North Pole. In comparison to sea temperature data obtained in 1987, which suggested that ocean water had decreased from 4°C in the ice-free Fram Strait to below 2°C between Svalbard and the Russian Islands of Severnaya Zemlya, the data for 1990 in the vicinity of Severnaya Zemlya was approximately 2.8°C. But they hasten to disavow any conclusion of a warming. They maintain that "the sparseness of the available data simply does not allow any such conclusion. . . . The year-to-year changes of 2°C could mask any trends in temperature development." Their data also clearly show that, over the period 1980–1990, ocean water temperature in that area varied regularly between 0°C and below to 6–7°C. Thus if a signal is to be found, and an early one at that, it will be necessary to carefully monitor this area. At this point, however, evidence for a warming in this highly sensitive area is not yet at hand (8).

Increasing sea temperature in the polar regions would be one signal, rising air temperatures could be another. Thus the recent report of disintegration of the Wordie Ice Shelf in Antarctica is of pressing concern. In the late 1970s J. H. Mercer of the Institute of Polar Studies, Ohio State University, contended that "a major disaster—a rapid 5 meter rise in sea level caused by deglaciation of West Antarctica—may be imminent or in progress after atmospheric CO_2 content has only doubled." He believed that doom was impending. "Although," he warned, "the models are known to be crude and oversimplified, so that the climate changes that will actually occur will no doubt differ considerably from their estimates, there is, at present, no way of knowing whether the models err on the optimistic or pessimistic side." But it was his following remark which is of more current concern. "If," he noted, "the CO_2 greenhouse effect is magnified in high latitudes (latitude 80°S) as now seems likely, deglaciation of West Antarctica would probably be the first disastrous result of continued fossil fuel consumption." And warming to his subject he continued as follows: "if the present highly simplified climatic models are even approximately correct, this deglaciation may be part of the price that must be paid in order to buy enough time for industrial civilization to make the changeover from fossil fuels to other sources of energy. If so, major dislocations in coastal cities, and submergence of low lying areas such as much of Florida and The Netherlands, lies ahead. . . . One of the warning signs that a dangerous warming trend is under way in Antarctica will be the breakup of ice shelves of both coasts of the Antarctic Peninsula, starting with the northernmost and extending gradually southward. These ice shelves should be regularly monitored by Landsat imagery" (9). And so they have. The recent Landsat images presented by Doake and Vaughn of the British Antarctic Survey take on considerable significance. However, according to these investigators the Wordie Ice Shelf has been breaking up over the past several decades. And it is, as H. Jay Zwally of the NASA/Goddard Space Flight Center tells us, "tempting to see this breakup as a sign of impending doom." He raises such germane questions as (10): "How should we view the new result in the context of public concern about greenhouse warming and estimates of rising sea level?" Both he and Doake and Vaughn believe that the impending breakup is a local phenomenon and "should not be extrapolated to other parts of the Antarctic" (11). Apparently temperature trends in most regions of Antarctica show no changing trends and sea ice extent, which lags temperature changes, also shows no shifting trend. Here too the investigators suggest that while substantial additional warming would be required before a breakup of the Ross, Ronne, and Filchner Shelves would even be considered, it is necessary to monitor this area carefully. It is also of the utmost importance to realize that ice shelves, unlike grounded ice, float in seawater, so that their breaking up into icebergs has no direct effect on sea level. But they are especially vulnerable to climatic change because of their exposure to both oceanic and atmospheric warming. Hence reports of sea and air temperatures in the polar region are of prime concern given recent model predictions of increased warming over the past 100 years and the continued emittance of CO_2 into the atmosphere. Nevertheless, whatever warming has occurred, it is not yet sufficient to have produced observable effects in either the sea, the air, or the ice.

COMPREHENSION OF WARMING EFFECTS

How can the warming effects of 0.5°C be comprehended? Perhaps the adventures of Erik the Red—Erik Thorvaldson—may be of assistance. Erik, son of Thorvald Asvaldsson, followed in his father's footsteps. The sagas tell us that in the year 982 Erik killed two men and was banished from his home in Iceland. Two years earlier his father Thorvald had killed a man in Norway and so was banished. With his family and a band of followers, Thorvald had sailed off and established a colony in Iceland. In his subsequent banishment Erik and his band headed west and landed on warm, rugged but protected shores with adequate land for farming. Conditions there were hospitable and the region green and fertile. He called the place Greenland. In fact, Erik had arrived in Greenland during a particularly warm period, the Medieval Optimum about 1000 years before the present. During this warm cycle the sea ice melted to such an extent that the Norsemen, the Vikings, not only settled Iceland and Greenland but reached Labrador and more than likely as far south as the coast of what is today the northeastern United States. It was also with the breakup of the sea ice that they sailed the great rivers of continental Europe as far east as Russia, and also reached the Mediterranean where they traded with the Italians and Arabs. Erik's two main settlements in Greenland prospered for almost 400 years. By the 13th century temperatures began to decline and the ice began to build and spread once again. By the 1450s the colonies were abandoned when the settlers died and others could not reach the island, as the Little Ice Age took hold. Over the period 900–1400 the temperatures were from 0.5 to 2.0°C warmer than today. Thus only relatively small temperature fluctuations can produce major climatic effects. Furthermore, during the Medieval Optimum, with its warmer temperatures, atmospheric carbon dioxide was at preindustrial levels. Is this an observation germane for the late 20th century?

CLOUDS AND THEIR INCLUSION IN MODELING

One of the greatest sources of discomfort to climate modelers has been and continues to be the appropriate representation of clouds in their models. Nevertheless, most all agree that clouds play a powerful role in regulating the earth's climate. It has long been known that clouds both cool and warm the earth; but with the recent availability of satellite-gathered data it has been found that clouds are far more powerful than heretofore assumed, and that they are so delicately balanced that temperature increases could easily disrupt the balance. More importantly perhaps, and this harkens back to filling in voids when data are absent, it has now been shown that clouds in the tropics differ from those in the temperate latitudes and each affects the total heat balance differently. This difference extends to clouds over the oceans and those over land masses. Thus models must at least be rerun to account for these differences. What this will do for model estimates of temperature increases remains yet another uncertainty.

One of the more important aspects of clouds that has been learned from the data obtained by the Earth Radiation Budget Experiment (ERBE) satellite was that clouds reflect more heat from the sun than formerly understood, exerting a net cooling effect. And the clouds over the tropics, which account for some 20% of the earth's surface, exert a disproportionate effect on total heat balance. Little of this information could have been included in models prior to 1990. Additionally, if temperature should increase, a warmer ocean would induce increased cloud formation, again increasing both the heating and cooling effects—a positive feedback loop effect. Accordingly, much rethinking will be required. Most recently, Robert D. Cess and 18 co-authors reviewed 14 atmospheric general circulation models for the quality of their interpretation of cloud feedback processes (12). The 14 models gave 14 different responses to cloud feedback processes, and they disagreed as well about magnitude and sign—whether the feedback was positive or negative. Between some models, differences in feedback effects ranged as high as 300%. Commenting on this, Anthony Slingo of the National Center for Atmospheric Research remarked that "it is becoming apparent that uncertainties in the treatment of clouds severely undermine model predictions of climate" (13).

An additional source of great potential uncertainty was advanced by Mitchell and co-workers of the Meteorological Office, London. They reported that their model indicated that changes of state of cloud water—its content of ice crystals—could substantially reduce the sensitivity of their model to increases in CO_2. Thus for their treatment of clouds there was a negative feedback in which a doubling of CO_2 resulted in a far smaller rise in global temperature. They concluded that "it is not possible to attribute the 0.5°C rise unambiguously to the effect of increases in trace gases" (14). To their credit a number of climatologists believe that poor data on clouds have introduced a fundamental error into all models and that it may be 10–15 years before new temperature estimates become available.

From time to time climatologists have posed the question of whether the earth's climate is self-regulating. That is, is there a natural mechanism that maintains the plant's global mean surface temperature close to 15°C (59°F)? If there is such a mechanism, the threat of global warming would evaporate. While such an idea would demand incontrovertible proof, as well it must, recent observations of the radiation budget and measurements of sea surface temperature in the tropical Pacific, by Ramanathan and Collins of the Scripps Institution of Oceanography, suggest that a natural thermostat exists that would prevent a runaway greenhouse. Although a "super greenhouse effect" is now known to occur during an El Niño event, it has also been observed that highly reflective dense clouds are also formed, which effectively shield the oceans from further heating. This indicates the operation of a natural, closely coupled ocean–atmosphere system, which would prevent warming from spiraling out of control. If the satellite radiation data obtained from the eastern and central Pacific can be shown to be generalizable to the entire planet, then indeed a climatic thermostat could set a ceiling on the level of warming that emissions of IR gases might engender. The idea is stunning in its simplicity and startling in its implication. Actually, Ramanathan and Collins found during the 1987 El Niño event that as sea surface temperature increased, water vapor also increased (as would be expected) and as did it absorbed more of the long-wave radiation, producing a super greenhouse effect. But as water vapor increased, there was simultaneous formation of large cumulonimbus clouds and at the tops of these clouds, ice crystals formed cirrus clouds in the cold troposphere. These thick, anvil-type clouds spread thousands of kilometers, and the more dense they became the greater was their reflectivity—more than enough to counterbalance the warming below. The net result is the prevention of ocean temperature from rising further. When cooling begins, the clouds disappear and the process begins anew. "What," they ask, "is the implication of this negative feedback if its validity extends to a perturbed atmosphere?" Their response will surely produce its own perturbation among climate researchers. For they maintain that "it would take more than an order of magnitude increase [10×] in atmospheric CO_2 to increase the maximum sea surface temperature by a few degrees, in spite of a

significant warming outside the equatorial regions. . . . In this regard,"
they continue, "the present hypothesis departs considerably from modern
general circulation models" (15).

Here again is a newly discovered climatic phenomenon, which models
have yet to incorporate. The searing question that remains to be answered
is: Does this process, which appears to operate when sunlight and sea
currents are the sources of heat, also operate when the sources of heat are
accumlating gases? Obviously, the need to determine this is of the highest
priority. How such an experiment would be undertaken is at least as
important. If a natural thermostat does in fact exist, and there are no
unforeseen countervailing and/or untoward side effects, climatic theory
may require reinterpretation.

THE OCEANS

The oceans are well understood to have a singular moderating effect on
temperature and pose yet an additional order of uncertainty. The vast
bodies of water are the most important sources of atmospheric moisture,
clouds, and heat. Being so vast, covering some 70% of the earth's surface to
depths of thousands of feet, their influence on temperature and its potential shifts should be substantial. Yet ocean data from around the world are
remarkably sparse. Temperature, salinity, and sea level are not all that well
documented, and their coverage over substantial periods of time and locations is spotty, requiring assumptions in place of data. A trenchant example is the wide difference in available data between oceans of the northern
and southern hemispheres. Southern hemisphere oceans are less well
known; they are also larger and on that basis alone take longer to heat up.
The thermal inertia of the ocean—its ability to store heat—is poorly
known, yet it must play a major role in trace gas-induced climate change.
Indeed, many scientists believe the depths of the oceans have not yet
adjusted to the current world climate, that they are still reacting to the
colder Little Ice Age of the 15th to mid-19th centuries. Here again there is
little past history of a quantitative nature to suggest how the ocean responds to climate forcing over centuries. Without this type of information,
verification of model predictions is all but impossible. And whether the
ocean and atmosphere are coupled or treated separately is another source
of uncertainty. Realistically, they should be treated as coupled, but models
and computers have thus far not been able to accommodate both. They may
in the future.

Differences in model predictions also arise from treatment of the oceans
in terms of type of circulation and whether thermal response time is
instantaneous or constant. An accurate thermal response time is essential
for detection of anthropogenic perturbations. In the Marshall Institute
Report, *Scientific Perspectives on the Greenhouse Problem* (16), it is noted

that when greenhouse gas emissions are projected to increase at the level of
1% per year, which corresponds with the Mauna Loa data, temperature
was predicted to rise in the next 50 years by 2–3°C over the entire earth, if
the effect of ocean currents is excluded. If currents are added to model
equations and the model rerun, only the northern hemisphere shows the
2–3°C increase. The southern hemisphere gains only 1°C, while the Antarctic barely warms at all, and may even cool slightly. As this has great
import for the degree of melting of the great ice sheets and by extension sea
level increases and potential coastal flooding, whether one or the other
treatment of the oceans is incorporated into model equations becomes
more than an academic exercise. Clearly, both cannot be correct, but both
could be unrealistic if other pertinent conditions have been omitted or
inadequately represented.

One of these is the time required for the ocean–atmosphere system to
respond and adjust to an abrupt change in temperature. Is it 10 or 100
years? Walter Munk of the Scripps Institute of Oceanography recently
noted that this would depend on ocean mixing processes, and he remarked
that "actual observations of atmospheric surface warming are based on
temperature time series at some 1000 locations. They suffer," he tells us,
"from the fact that many of the land stations are contaminated by so-called
'urban heat islands' where the microclimate has undergone significant
changes. Sea surface temperatures are contaminated by a systematic conversion from bucket to injection measurements; the bias so introduced may
constitute as much as 50% of the observed change since the turn of the
century." And after attempting to correct for the biases, various estimates
suggest a surface warming (atmosphere and ocean) of 0.3–0.7°C. Quite
aside from these measurement problems, Barnett and Schlesinger have
discussed the difficulty of detecting the global greenhouse warming
against the background of the inherent air temperature variability. They
find that greenhouse warming and inherent variability have similar spatial
patterns, thus suggesting that the air temperature field is a difficult place to
attempt early detection of the greenhouse signal. By contrast, they suggest
that the oceans may be a better environment for early detection of a greenhouse signal (17). This idea has not been lost on oceanographers—
especially Munk. He has devised an elegant but relatively simple procedure for taking the temperature of all the oceans simultaneously. The idea
is to measure the speed of sound waves traveling a known distance through
water. The time it takes to travel thousands of miles depends on the density
and pressure and hence temperature of the water. The technique is in effect
a global ocean thermometer and is referred to as ocean acoustic tomogra-
phy (OAT); it is not all that different from the type of imaging produced by
medical CAT scans. In its early tests it consisted of three transceivers, echo
chambers (Figure 8) anchored in a triangular pattern on the ocean floor
1000 kilometers (600 miles) apart and approximately 700–750 meters
below the surface. Each unit receives and sends the signals, consisting of

Figure 8. Echo chamber used to transmit and receive sound waves sent through the oceans as a means of measuring water temperature. (Courtesy of W. H. Munk, Scripps Institution of Oceanography, La Jolla, CA.)

low hums produced by bending an aluminum sheet back and forth by a hydraulically driven piston. The humming sound travels at the rate of 1.5 kilometers per second and at its frequency of 250 hertz covers the 1000 kilometers in approximately 11 minutes. Having passed its preliminary tests in Hawaii, the plan now is to locate a permanent sound source in the South Indian Ocean off Heard Island, an Australian possession, from which sound will be emitted in 80-second bursts every 2 hours for a 24-hour period every fourth day. This time the frequency will be reduced to 50–60 hertz, which will produce less "noise" interference from ships and other emitting sources. Additionally, accoustic receivers will be established, as shown in Figure 9, at multiple sites on each of the six continents. This network of transceivers is expected to detect changes of density and temperature quickly and precisely, in fact, to hundredths of a degree (18).

If global warming proceeds as predicted, the seas should warm in parallel, and as sound is transmitted, it should increase in speed from point to

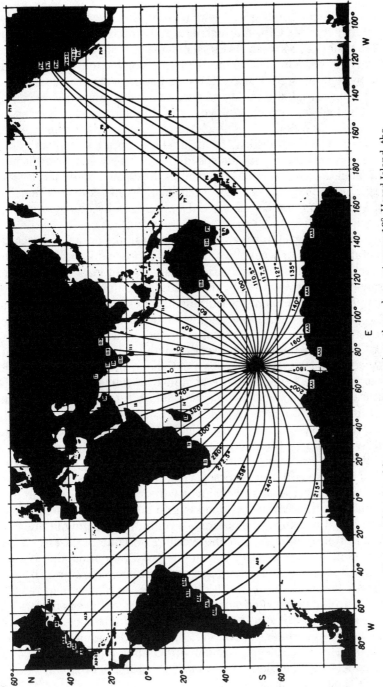

Figure 9. The curved lines represent acoustic paths and are set every 10°. Heard Island, the site of the sound source, is located at the convergence of these paths. The receiver sets are noted by circles. (Courtesy of W. H. Munk, Scripps Institution of Oceanography, La Jolla, CA.)

point with each unit increase in temperature, since warmer water decreases in density and with that decrease, the speed of sound must increase. The distances between Heard Island and the microphone receivers are precisely known. Thus careful measurements of the time taken to reach any receiving station (sensor) will allow calculation of the average speed of the sound from initial site to destination. In water the average speed of sound is 5280 feet per second (1760 yd/sec). Although speed is affected by salinity, current direction, and compression of the water with depth, water temperature is the most critical factor affecting speed of transmission. Accordingly, from year to year the speed of sound between Heard Island and San Francisco, a distance of 11,210 miles (6720 km) should increase—traversing the distance in less time each year. Munk estimates that the speed of sound over the course should be reduced by one-quarter of a second each year.

Simply suspending a traditional thermometer in the ocean and recording temperature daily or weekly would be subject to confounding, due to normal background temperature variation, and would mask changes due to excessive warming within the "greenhouse." Confounding could only be overcome by thousands of measurements taken over 50 or 100 years. The OAT procedure permits averaging-out the background "noise" by taking several measurements over a brief period.

Currently, tests are underway to ascertain if the signal from Heard Island can be detected as far away as the coast of Canada and Coos Bay, Oregon. Preliminary data are expected to be obtained by 1992. When these new temperature values become available, modelers will have firm data for the first time to insert into their models, which will further increase their validity. Over the past 20 years the available ocean temperature data were obtained by dangling a thermometer over the ship's side for weeks at a time. By the time readings were obtained, the oceans had changed. The new OAT technique will provide precise readings almost instantly. And to transmit the low level of sound planned without losing energy through outward dispersion, it will be confined within a "sound axis"—a layer of water extending through all the oceans—a conduit for sound but one that exists at differing depths around the world. Consequently, transmitters and receivers must be positioned at a number of different depths. Within the sound axis sound is transmitted at a constant rate; below and above it sound moves faster. Sound moving through the "axis" tends to remain within it. Here then is an experiment with a prediction that can be verified relatively easily and well within the time frame useful to decisionmakers. The preliminary experiment to test the concept and equipment began in January 1991. The full scale experiment will continue to 2001.

Some models indicate that the temperature of the oceans may rise as much as 0.1°C per year. Oceanographers, on the other hand, do not believe that atmospheric physicists running their models have the type and sufficiency of data about the oceans necessary to make appropriate predictions.

Thus the OAT experiments could be of great benefit in providing accurate information about the temperature of deep waters heretofore unavailable to modelers.

The OAT project is imbued with even greater urgency and meaning given the problems of bias in existing temperature data sets. If we are to invest credence in the 0.5–0.6°C temperature increase suggested to have occurred since 1860, then these few degrees must be without blemish. According to Thomas R. Karl of NOAA and Philip Jones of the University of East Anglia, "all global (land-based and marine) area-averaged temperature data sets are contaminated by a number of biases of varying magnitudes . . . . All . . . land-based . . . data sets suffer from a number of biases of which the most serious may be the urban-warming bias. A number of large biases (several tenths of a °C) also exist in the marine data." And they inform us also that "the potential bias of the urban effect in global data sets has not been definitively assessed. At present, only rough estimates of the potential impacts of urbanization can be given. This includes an urban bias in the Hansen and Lebedeff (1987) data over the United States between 0.3 and 0.4°C over the twentieth century which is larger than the overall trend in the United States during this period . . . . "To our knowledge," they go on to say, "the United States is the only large area of the globe where the magnitude of this bias has been thoroughly studied" (19). Thus the almost instantaneous seawater temperatures to be obtained by Munk's underwater broadcasting system must be seen as of inestimable value, both to increase model accuracy and to test the data on which so much forecasting rests.

Similarly, the World Ocean Circulation Experiment (WOCE) has been organized to obtain circulation data at moments in time between 1991 and 1996. For the past 5 years a related program, TOGA (Tropical Oceans and Global Atmosphere), has been gathering circulation data in a restricted area; but both of these are in troubled financial straits. WOCE may not obtain the funds to begin its work, and TOGA may be unable to complete the final 5 years of its data gathering. Both are subject to a depressing form of bureaucratic double think in which the scientific data underpinning warming are vital and needed, and their unavailability stalls decisionmaking; yet without a supporting base of collected data, funding is either unavailable or available at levels assuring that research efforts will be all but impossible to pursue. President Bush indicates that he prefers to avoid precipitous moves on global warming given the uncertainty of the predictions, yet funds are withheld from projects that could supply the data.

Internationally, scientists are reasonably well agreed that a global climate observation system, similar to what World Weather Watch does for the atmosphere, is needed for the oceans. Five to six billion dollars would be needed to fund a joint worldwide effort, but hopes of obtaining it are slim, especially as a recession/depression spreads.

CHANGES IN OCEAN TEMPERATURE AND ITS EFFECTS
ON SEA LIFE

The Atlantic Ocean is snake free. The Pacific and Indian Oceans are not. The Atlantic is far colder and has been separated from the Pacific for 3 million years—ever since the Isthmus of Panama rose from the depths and joined North and South America. The coldness of the Atlantic and the physical barrier separating the Atlantic and Pacific have kept sea snakes out of the Atlantic.

Sea snakes are the most abundant of the earth's reptiles and are more venomous than cobras, from whom they descend. *Pelamis platurus*, the yellow-bellied sea snake, plies the warm waters of the Pacific and Indian Oceans where understandably it has few predators. Most potential threats have learned to give these bright yellow and black specimens a wide berth, their small size notwithstanding. For a snake of less than 3 feet in length and under one-half pound in weight it obtains wide respect, primarily because milligram for milligram its venom is deadlier than a cobra's. *Pelamis* lives out its several years in seawater no lower than 18°C (65.4°F). If sea temperatures rise and currents shift, then *Pelamis* could well break out of its restrictive habitat. The appearance of a few venturesome members of the genus *Pelamis* off the coasts of Boston or Baltimore would signal a sea change in temperature, perhaps well before physical measurements; or would verify measurements made by Munk and co-workers.

Given the rapidity with which new data on the ocean become available, two further comments are appropriate. At the Geophysical Fluid Dynamics Laboratory, rather than programming their supercomputers to suddenly "inject" a double CO_2 concentration, their equations called for a 1% increase from 1958, the point at which CO_2 levels began being collected at Mauna Loa. After a simulated "run" of 30 years, global surface temperature had risen 0.7°C. After another 20-year run, for a total of 50, surface temperature had increased by 1.4°C; and after an additional 20-year run, for a final total of 70 years corresponding to the year 2028, and a doubling of atmospheric CO_2 levels, simulated air temperatures were found to have risen by 2.1°C (3.8°F). But most surprising and wholly unexpected, their data showed that the Antarctic Ocean had not warmed as anticipated. Perhaps the atmosphere–ocean system handles sudden emissions differently than similar emissions over the longer term. That remains to be seen. But it does suggest that knowledge of the oceans' use of CO_2 is far from complete.

Further confounding an anticipated warming may be continued El Niño events. It is now well established that the repressively hot summer of 1988 was the result of the 1986–1987 El Niño, not a signal of global warming. Before continuing, a brief digression to explain the El Niño phenomenon seems in order.

El Niño is a child of a fickle ocean. Normally, the upwelling nutrient-rich cool Pacific current flows northward along the coasts of Peru and

Equador, supporting a large fishing industry. But every few years the upwelling ceases; the current becomes warm and flows southward. Rainfall increases dramatically, bringing flooding to coastal areas. Without upwelling there is a loss of nutrients for a broad range of marine species, which produces serious economic dislocation for the fishing and related industries.

As Quinn and Neal recently reported (20), El Niño events have been well documented over the past 450 years, while others have found evidence of their occurrence back to 500 B.C. Clearly, they are yet another climatic phenomenon with which humankind must contend. Normally, the rains accompanying the currents, coming as they do in late December, are so regular and so welcome in these otherwise arid lands that they have come to be called "El Niño," the Christ Child—sustaining life throughout the many months of drought. In 1982–1983, after a decade of normal weather, El Niño changed abruptly, bringing with it the strongest warming of the equatorial Pacific in this century. The warm current brought coastal Equador and Peru close to 600 mm of rain during June, a "dry-season" month in which normal rainfall barely manages 10 mm. The unseasonably warm weather front stretched 13,000 kilometers from the coast of South America to the archipelagos of the western Pacific, creating a drought there that enveloped Australia. Over a thousand people died and staggering economic damage occurred in Australia, Asia, South America, and Africa. This havoc resulted from an elevation in sea temperature of less than 5°C. Figures 10a and 10b compare normal atmosphere–ocean conditions with an atypical El Niño pattern. Four years later it struck again. This time its large-scale shifts in ocean temperatures disrupted climate in North America, India, and Bangladesh, among others. In the United States and India, it was responsible for 1988s searing summer heat and thousands of heat-related deaths. It is now well established that it was El Niño, or more specifically La Niña, which brought sweltering weather to Washington, DC during the Congressional Hearings in which James Hansen assured Congress and the media that we were reaping the bitter harvest of excessive use of fossil fuel and that its consequence—global warming—was at hand. The admonition was premature. But it focused the mind as never before.

El Niño is now known to have a twin, a cold phase dubbed La Niña; and it appears to have been the cold phase which brought the repressive heat of 1988. While it remains to be determined what triggers both the warm and cold phases, El Niño/La Niña remain elements of climate to be reckoned with—so much so that a substantial body of scientific opinion holds that the warm and cold swings in the eastern Pacific, because of its vastness, have more effect on global climate changes than accumulation of atmospheric gases. This remains to be established.

Scientists at the Shangai Observatory of the Chinese Academy of Sciences believe that the anomalous sea-surface temperature of 1986–1987 induced by El Niño has been observed once again, and they predicted that

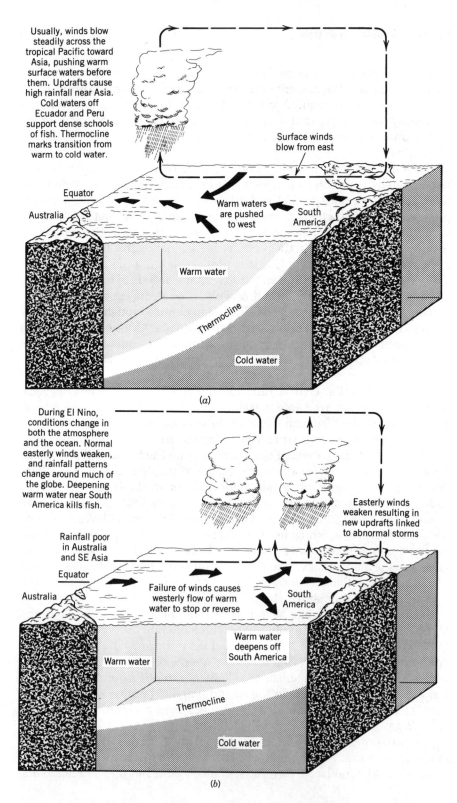

Usually, winds blow steadily across the tropical Pacific toward Asia, pushing warm surface waters before them. Updrafts cause high rainfall near Asia. Cold waters off Ecuador and Peru support dense schools of fish. Thermocline marks transition from warm to cold water.

Surface winds blow from east

Equator

Australia

Warm waters are pushed to west

South America

Warm water

Thermocline

Cold water

(a)

During El Nino, conditions change in both the atmosphere and the ocean. Normal easterly winds weaken, and rainfall patterns change around much of the globe. Deepening warm water near South America kills fish.

Easterly winds weaken resulting in new updrafts linked to abnormal storms

Rainfall poor in Australia and SE Asia

Equator

Australia

Failure of winds causes westerly flow of warm water to stop or reverse

South America

Warm water deepens off South America

Warm water

Thermocline

Cold water

(b)

it would reach peak intensity by mid-1990 and last into 1991. Its effects are predicted to be greater than the last one. The Chinese researchers Zheng, Song, and Luo have published data showing a close if not direct relationship between variation of the earth's rotation and development of El Niño events (21). They have shown that changes in day length are related to these events. For example, their data indicate that the interannual rate of the earth's rotation decelerates when the seas warm up and accelerates when they cool. "Therefore," they say, "every El Niño event usually occurs after the turn of the interannual rate of the earth's rotation from acceleration to deceleration." In addition, they found the amplitude of the change to be correlated to the strength of an El Niño, and they maintain that the time and magnitude of an El Niño can now be predicted. Should their calculations be correct, the summer of 1991 could be breathtakingly warm, but most importantly it will have to be differentiated from warming due to accumulation of greenhouse gases. This will surely be closely observed because the early years of the 1990s have been designated by climate modelers as the time warming signals should be expected. Thus the need to differentiate and disentangle these two climatic phenomena becomes imperative. It will also be instructive to see how these scientists fare as forecasters.*

Most recently the discovery of massive blooms of the algae Phaeocystis across large expanses of the Greenland Sea indicates that the northern hemisphere and its oceans may be a far larger sink of atmospheric CO_2 than heretofore understood. Thus, the accumualation of new data suggests greater importance of the northern hemispheric region for the global carbon cycle, and calls into question the data previously used in model scenarios.

Although the exchange of carbon between the atmosphere and biosphere is constant, the rate and magnitude of exchange is poorly known. It has been a tenet of atmospheric chemistry that the oceans of the southern hemisphere were the major sinks for atmospheric CO_2, while land plants were the major repository in the northern hemisphere. Thus the recent report by Walker O. Smith, Jr., of the University of Tennessee, along with six collegues from universities across the country calls this into question. It may well be that phytoplankton activity in the arctic is a major participant in the global carbon cycle providing a previously unknown carbon sink (22).

*They have proved to be as good as their data. The summer of 1991 was one of the warmest of the 20th century.

Figure 10. (a) Characterization of the normal climatic conditions prevailing off the coasts of Equador and Peru. (b) The severely altered conditions of an El Niño event. (Reproduced with permission from Melvin A. Benarde, *Our Precarious Habitat: Fiften years later*, John Wiley & Sons, New York.)

Similarly, Kirchman of the College of Marine Studies, University of Delaware, and co-workers, recently found the turnover rates of dissolved oceanic organic carbon during a spring phytoplankton bloom, in the north atlantic to be greater than previously reported (23). Again, if this new data is found to be widely generalizable, it would "seem to demand changes in models of carbon cycling and of the oceans role in buffering increases in atmospheric CO_2."

Perhaps at this time forecasting ought to be looked on as an exercise for gaining experience rather than as a tool for decisionmaking.

CARBON DIOXIDE CONTENT OF ATMOSPHERE

Carbon dioxide, the trace gas of central concern to greenhouse warming, is itself a source of uncertainty. Consider that the upward trend in atmospheric CO_2 is evident at all stations where it is monitored. Nevertheless, when the increases are translated to tons of CO_2 and compared with tons of fossil fuel known to have been combusted with CO_2 emitted, no more than 55–58% of the CO_2 produced can be accounted for in the atmosphere. Some 40–45% is missing. Given this conundrum, climatologists such as Charles Keeling have postulated that reasonably it should be found in the oceans, where uptake is probably far greater than originally believed. This idea was well received and appeared to satisfy many concerned scientists—until recently, when Tans, Fung, and Takahashi published their model calculations, which predicted that the missing CO_2 would be found in the boreal forests of the northern hemisphere (24). The shock waves set off by that report are still reverberating within the scientific community, where there is now greater concern over whether their knowledge of the highly complex carbon cycle is as strong as they believed it to be. Uncertainty about the actual movement of carbon, its sinks and sources, must create additional difficulties for modelers and their models. A problem that arises is that models yield numbers, and numbers are invested with a reality difficult to dislodge.

Of overriding importance for estimation of trends is knowledge of past CO_2 levels. That atmospheric gases undergo natural variations irrespective of human activity is now well established. During the last ice age, bubbles of gas trapped in the ice indicate atmospheric CO_2 levels of 210–220 ppm, fully 66% of current levels, and during the preindustrial period—up to the 1840s—levels of CO_2 have variously been determined as anywhere from 250 to 290 ppm. Depending on the figure used, an error on the high side of as much as 0.3 watt per square meter for current calculations can occur.

In discussing the problems of carbon dioxide emissions as a consequence of deforestation, Detweiler and Hall maintain that people the world over, with their ever growing needs for land and timber, assure that forests will continue to be devastated, while the CO_2 released by these activities will affect the carbon cycle and hence climate change. Yet they conclude that "it is remarkable that our quantitative estimates of tropical land use change and its consequences are so uncertain" (25). Thus predicting CO_2 levels in the year 2040 or 2060 requires a degree of comprehension of the biogeochemical carbon cycle not yet available, except possibly in broad outlines, as well as considerable vision as to future economic, agricultural, and deforestation activity—the types of assumptions responsible for model uncertainty. It is therefore quite surprising that so many model predictions fall within a narrow range. Perhaps this may be a point on which to focus.

In addition to trace gases, airborne particles—aerosols—and sunspot activity appear to affect insolation and thereby the earth's thermal balance. Some models have attempted to deal with these effects; others have not, as their treatment can be far more complex than CO_2.

AEROSOLS

Particulate matter can occur naturally via volcanic eruptions and artificially via industrial stack gas releases, as well as such agricultural practices as deforestation (with burning) and land clearance for farming, during which dry soils become airbone by local winds. Thus airborne particles or aerosols consist of dust, soot, ash, volcanic emissions, sea salt, and products of the decay of plants and animals and other artificial and material debris. Increases in these can produce both warming and cooling effects depending on a range of factors including altitude, size composition, and spatial and temporal distribution of the aerosol layer, as well as the characteristics of the underlying surface—its albedo. Suspended over darker surfaces such as oceans, with their low reflectivity, the effect tends to shield the earth from incoming solar radiation, hence exerting a cooling effect. Over land masses such as deserts or snow-covered areas with high albedo, they exert a trapping effect that produces a net warming. The balance between shielding and trapping effects can determine the particles' overall thermal influence. As shown in Figure 11, particles can scatter a portion of the incident radiation back into space and can also prevent its passage earthward, thereby directly modifying the atmospheric radiation budget. In fact, it has been shown that increased aerosol loading can perturb the radiation budget as much as a doubling of carbon dioxide. But it can also produce cooling effects. Thus the correct treatment of aerosols in models assumes increasing importance and may even be pivotal. But it remains to be ascertained just what that correct treatment is.

An excellent example of this "one way or another" effect was recently presented by T. M. L. Wigley of the Climatic Research Unit, University of East Anglia, and one of the foremost practitioners of climate modeling. Wigley's new data will further complicate decisionmaking. It is well

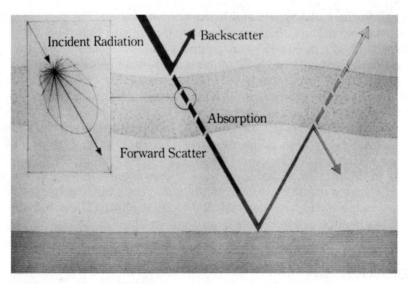

Figure 11. Radiation scattering exhibited by a layer of particles. The inset shows the distribution of scattering by a single particle of mean size.

known that, in addition to emitting CO_2, burning fossil fuel also liberates sulfur dioxide, a source of particulate matter. These SO_2 particles reflect sunlight back into space, which cools the earth and tends to offset warming. That is, the presence of CO_2 and SO_2 can offset one another's effects. This phenomenon was described almost two decades ago but was given little attention. Wigley has resurrected this physicochemical effect and invested it with new respect. Thus he informs us that any precipitous reductions in burning fuels would reduce particulate matter and further exacerbate warming for as much as 30 years. Furthermore, without benefit of an aerosol effect, it would be well into the 21st century before any noticeable benefit of CO_2 reduction would take effect, because of the time required for the carbon cycle to absorb it. Given the importance of his new data, let us hear from him directly. "Even with no aerosol effect, it is several decades before any noticeable radiative-forcing response to an emissions reduction policy occurs, simply because the carbon cycle has such a long response time. The aerosol effect delays the response still further, completely offsetting it for up to three decades." The full impact of this will be a while in coming. It must of necessity be seen as a setback to those who seek fossil fuel reductions now. But those who prefer to wait can find no ally in Wigley, who realized full well how his data would be received by those who prefer "business as usual." "It would be entirely wrong," he cautioned, "to consider SO_2-related cooling in isolation as a benefit . . . . CO_2 forcing could be markedly offset by the effect of sulphate-aerosol increases, but continued growth in emissions would lead to a rapidly widening contrast in inter-hemispheric forcing, potentially even more disruptive to the climate system than a uniformly distributed 'pure' greenhouse effect'' (26). He was clear in warning that advocates of watchful waiting not see the SO_2 effect as a panacea. But many will, if only because it opens the uncertainty box ever wider. What new findings await us a month or a year from now that could further modify previous forecasts? Given that few if any model programs have included this aerosol effect, it could become a pivotal confounding factor. It will require much attention.

Volcanic eruptions further muddle projections of greenhouse warming. It is well known that large eruptions such as Tambora and Krakatoa in Indonesia had decided cooling effects over large areas of the world. For example, in 1816 in New England it snowed in June, with killing frosts through August. That was the year without a summer—the consequence of the eruption of Mt. Tambora in 1815. That eruption ejected some 80 cubic kilometers of debris into the atmosphere, which appears to have caused a drop in temperature of 0.4–0.7°C in the northern hemisphere. Whether smaller eruptions can effect a cooling is a source of current brisk scientific discussion. That Benjamin Franklin wondered if the unusually cold winter and cool summer of 1783 was due to the eruption of Hecla in Iceland in 1782 testifies to the continuing concern about volcanoes and climate.

The character of the dense debris hurled into the atmosphere by erupting volcanoes is shown in Figure 12. This is an aerial photograph of Mount St. Helens from the west, taken shortly after the May 18, 1980 eruption. USGS scientists estimate that the minimum volume of ash and rock ejected during the May 18 eruption amounted to about 1 cubic kilometer (1.3 billion cubic yards) of material. The estimate is based on an assumption that the new crater formed by the blast at Mount St. Helens measures about 1 kilometer wide, 2 kilometers long, and at least 0.3 kilometers deep.

If the May 18 eruption did produce as much as estimated, the volume would be about equal to the estimated volume of volcanic materials ejected by Mount St. Helens in A.D. 1500, and about one-third the volume of the volcano's 1900 B.C. eruption. By comparison, the eruption of Mount Vesuvius that buried Pompeii in A.D. 79 produced slightly more than 1 cubic kilometer of ash; the deadly eruption of Indonesia's Krakatoa in 1883 produced about 20 cubic kilometers of ash; and the largest eruption in history. Tambora in Indonesia in 1815, produced nearly 80 cubic kilometers of ash and other ejected material.

The May 18 eruption was the biggest of the series of eruptions that began on March 17, 1980. Its eruptive plume reached over 60,000 feet into the atmosphere, and its initial burst released the energy equivalent of about 10–50 megatons of TNT. As a comparison, the largest nuclear bomb test, conducted in the Soviet Union in October 1961, released the energy equivalent of about 50 megatons of TNT. Quite obviously, volcanoes are among nature's greatest forces and must be reckoned with in all climate models. But how are they to be properly represented?

Figure 12. Aerial view of the eruption of Mount St. Helens on Sunday morning, May 18, 1980. Photo was taken from the west. (Courtesy of the U.S. Geological Survey, Department of the Interior, Washington, DC.)

That sulfur dioxide is one of the major ingredients of volcanic debris is well known; that it may be the compound exerting the greatest cooling effect remains to be determined. We do know that it rises into the stratosphere where it is converted into droplets of sulfuric acid. The acid droplets have a shiny, highly reflective glasslike surface and thus are highly capable of deflecting insolation back into space—away from the earth—with a resulting cooling effect. And there appears to be evidence that volcanic gases, especially as thinly dispersed sulfuric acid hazes or veils, produce a cooling effect. But there is also evidence that thick dust clouds exert an opposite effect, producing a warming. Consequently, interpretations of radiative effects of aerosols on climate must be tempered by robust optical transparency data. Thick or thin layers could mean the difference between tolerable and intolerable conditions for humankind.

If the recent projections of Latham and Smith of the University of Manchester (England) are credible, aerosols of salt particles (sodium chloride) whipped up from the oceans may be sufficient to counteract the predicted warming. Latham and Smith are convinced that global warming may add sufficient additional energy into the atmosphere, making winds blow faster over the oceans, and so disturbing the surface that more salt particles would be lifted into the air, where they would create the nuclei around which water vapor would condense. This condensation would form large clouds, which in turn would reflect a larger amount of heat away from the earth and back into space. They maintain that the increased warming of 2°F predicted by the year 2025 would increase windspeed by 10–20 miles per hour, which would raise enough salty aerosol to fully compensate for the warming. And they take their predictions a step further. They believe a controlled field trial in which cloud-condensation nuclei are inoculated into the air to ascertain both cloud formation dimension and degree of heat trapping is not only appropriate but feasible. Given its relative simplicity and potential for shielding the earth, this type of experiment should be given the highest priority, which means ample funding, as it can supply the type of verification few model predictions can deliver (27).

Before the war in the Gulf, warnings were raised that smoke from oil fires would emit such large quantities of gases and soot that temperatures would drop, producing a form of "nuclear winter." As Kuwait was about to be liberated, hundreds of wellheads were set ablaze of retreating Iraqi troops, and smoke from burning oil has since blanketed the skies in the area. In March 1991, British scientists of London's Meteorological Office flew through the dense smog and found that although the skies over Kuwait, Iran, and Iraq contained the thickest smoke, low levels of soot were readily detectible well north of Turkey over the Black Sea, as far west as Egypt, the Sudan, and Ethiopia, and as far east as India and China. For a number of scientists, the small size of the soot particles are a cause of concern, if not alarm. They believe that the smallest of the particles can continue to rise into the stratosphere, remain there for extended periods, and block out the sun. Should that occur, and the wells continue to burn for 2–3 years—within the realm of possibility—and if as little as 1% of the soot reaches the stratosphere, there could be a drop in temperature of 1°C or perhaps 2°C across the northern hemisphere. If larger amounts rise, temperature reduction could approach that of a massive volcanic eruption, which would further confound and forestall the appearance of a warming signal. This is not the way to "buy time," as both carbon dioxide and nitrous oxides are being emitted into the atmosphere simultaneously. Everyone's best interests would be served by early quenching of the fires.

Mount Pinatubo is yet another potential confounder. The recent eruption of Mt Pinatubo in the Philippines is, by far, the largest volcanic eruption of the 20th century, and its outpouring of ash and aerosol material could indeed depress global temperature, further frustrating and confounding efforts to detect a warming signal. The Total Ozone Mapping Spectrophotometer—TOMS—aboard the *Nimbus 7* satellite has already registered between 15 and 16 million tons of SO_2—double that emitted by El Chichon in 1982. The SO_2 reacts with water molecules to form minute mirrorlike droplets of sulfuric acid, which act as reflecting surfaces for sunlight, turning the rays back into space. Unfortunately, these droplets can persist for years. And these droplets are different from those arising in Kuwait. The aerosol particles from Mt. Pinatubo are rising into the stratosphere while those generated by the oil fires appear not to be penetrating beyond the troposphere, where they also react with moisture but fall out within a few days. Additionally, the smoke plumes rising in the Persian Gulf are not as black as first believed; thus they do not have the reflecting power originally anticipated. Patently, until more is definitely known, a clear warming signal continues to remain elusive.

SOLAR ACTIVITY

There is yet another element that may or may not create additional uncertainty for model forecasts. Thus far solar output—luminosity—has not been considered variable. That is, modelers and climatologists do not hold with the idea of a variable sun. Consequently, general circulation models have assumed a constant sun. If the sun is in fact variable in its output over century-long periods, then natural processes could account for the 0.3–0.5°C temperature increase estimated to have occurred over the past 100 years.

As the sun is the earth's source of energy, it does not violate reason to believe that variation in its output could affect climate. Geologic evidence suggests that the sun's luminosity has not varied more than a few tenths of a degree over the past 3.5 billion years, and satellite measurements tend to support this. The key may just be tenths of a degree. There is good agreement among climatologists that a change of 1% in the amount of heat reaching the earth could shift the average global temperature by 1°C. Current theory, however, holds that the solar energy received at the top of the atmosphere on decadal time scales is too small to disturb the earth's total energy balance, and as such there is adequate reason to assume a constant sun in model programs. On the other hand, recent studies suggest this to be an untenable assumption. Just what has been learned to warrant or at least consider model reprogramming?

If one astronomical observation can be said to be amply documented, it is the appearance and disappearance of sunspots. They have been continuously viewed and recorded by the unaided eye since early Egyptian times, and by telescope since Galileo first trained his new instrument on the sun almost 380 years ago. He saw them, but he didn't see many. Through a telescope sunspots and flares (Figure 13) appear as dark blotches on a bright disk. These darker areas are also cooler areas, approximately 4200°C, compared to 6000°C at the sun's surface. It is now known that these highly active regions are associated with locally intense magnetic fields. The solar flare in the figure, produced by a sunspot, is a sudden eruption of

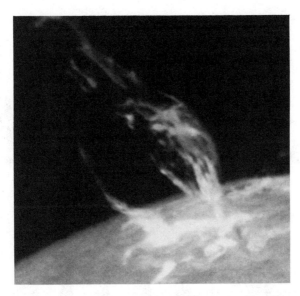

Figure 13. In an area of sunspot activity, this flare with associated eruptive prominence and spray was observed on June 20, 1989. The flare is seen being hurled some 150,000 miles beyond the surface of the sun. (Courtesy of the U.S. Air Force/Air Weather Service.)

energy shown here extending some 142,000 miles above the surface of the sun.

The idea that the sun is a steady source of warmth and brightness came into question in 1843 when Heinrich S. Schwabe, a German pharmacist and amateur astronomer, noted that the number of sunspots varied with a regular 10-year periodicity. Within a few years others refined his observation to obtain an 11-year cycle of appearance and disappearance. But as Figure 14 shows, although it has occurred without interruption between 1700 and 1990, the cycle is regular neither in frequency nor intensity—its amplitude ranging from a low of 45 in the early 1800s to a high of close to 200 in 1957. The current cycle, the 22nd, was estimated to have reached its peak in February 1990. By January it had surpassed the activity of the previous two cycles and may exceed cycle 19, the highest ever recorded. This is worth bearing in mind.

In 1894, E. Walter Maunder of London's Royal Greenwich Observatory published a seminal paper that created little if any interest. He attempted to call attention to a curious phenomenon—the almost total lack of sunspots in the 60 years between 1645 and 1705 (Figure 15), which, by the way, would explain why Galileo and those who followed him early on reported so few. Maunder was not the first to notice this unique astronomical event. Gustave Spörer, a German astronomer, had remarked on this sunspot-deficient period years earlier. Few seemed to be listening. This void has come to be known as the Maunder or Spörer minimum, but most

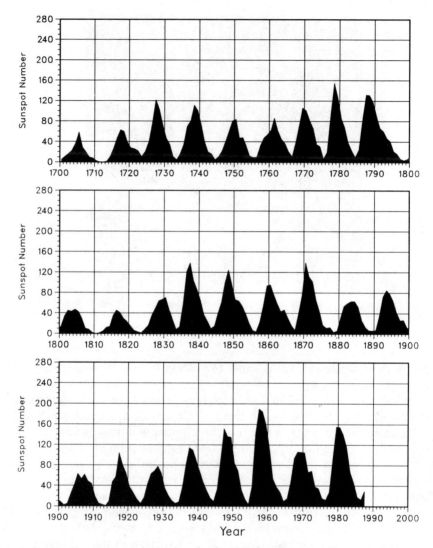

Figure 14. Yearly mean sunspot numbers for the period 1700 to 1987. (Courtesy of the World Data Center, NCAR, Boulder, CO.)

importantly it appears to have corresponded with the period of the Little Ice Age, which lasted from the 15th to the mid-19th century. It corresponded, but were the two related? If they were, it would be a momentous observation and highly significant for climatic conditions on earth. If a lack of sunspot activity coincided with a cold period, and if intense activity such as the decade of the 1980s as well as the increased activity since the turn of the century coincided with rising temperature, a natural process would explain the rise. The issue then is whether sunspot activity is related to climate and whether the sun is a variable star.

Additional evidence supporting the Maunder minimum is provided by the aurora borealis, the "northern lights." The sun emits a constant stream of high-speed, electrically charged particles—the solar winds—which enter the top of the earth's atmosphere where protons and electrons are in constant motion. At times of increased sunspot activity, the density of the solar wind increases and overloads the atmosphere causing—at heights of 100–1000 kilometers (65–650 miles)—charged particles to spill over into the upper atmosphere where they react with particles of air, inducing them to emit radiation that is seen as green, blue, or red auroras, the northern lights. (Actually borealis means northern and aurora means dawn.) When solar activity is particularly strong, these lights take on the shapes of bands, crowns, rays, or arcs. The key point here is its relationship to solar activity. Auroras are seen most frequently at higher latitudes, closer to the Arctic Circle, because the earth's mangetic field acts as a shield, protecting the atmosphere against incoming solar particles. And normally 5–15 displays of the northern lights have been reported from northern Europe annually. However, between 1645 and 1715 few, if any, were observed. From 1679 to 1715 none was seen. John A. Eddy, director of NCAR's Office for Interdisciplinary Studies, informs us that Edmund Halley of comet fame was 60 years old in 1716 when he saw and reported on an aurora, the first he had ever seen; but most important, he had watched for one all his professional life. "Although he didn't know it," Eddy wrote, "Halley's life had spanned most of the period of the Maunder minimum" (28). With the passing of the Maunder minimum and for the past 260 years, auroras and the northern lights have become a common feature of the earth's astronomical bag of tricks. But the association was less than direct. Needed were more concrete data. Trees, very old trees, possess that type of evi-

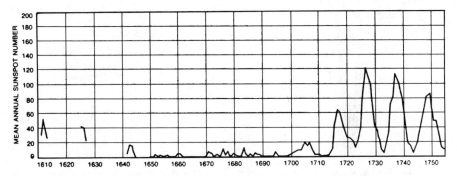

Figure 15. Plot of sunspot activity over the period 1610–1750. The years up to 1720 cover the period of the Maunder minimum.

dence. Variations in tree-ring widths from one year to the next are re-congnized as significant sources of chronological and climatic informa-tion. Studies of tree rings as a potential source of paleoclimatic data began in the early 18th century, and dendroclimatology, the use of treerings as an indicator of climate, has emerged as a well established scientific disci-pline.

A cross section of most temperate zone trees will show, as seen in Figure 16, an alteration of continuous lighter and darker bands around its circum-ference. These are seasonal increments produced by meristematic tissue, the microscopically thin layer, the cambium, shown in the figure. As the cambium cells divide, the tree grows in diameter. The new cells formed toward the inside of the tree become wood or xylem while the cells on the outside become the phloem layer. Figure 17 provides a concise overview of plant anatomy, histology, and physiology, with the location of the xylem and phloem shown in exploded view.

Cells produced in the spring growing season are larger than those pro-duced in the summer, which because of their smaller size and density appear dark. It is the alternation of dark and light areas which yields the rings. In the world's temperate zone all trees add a ring a year. The section of the trunk shown in Figure 16 has a series of 10 alternating dark/light annual rings, each pair corresponding to a year's growth, and hence the tree is 12 years old. The rings in Figure 18 indicate a tree that is approxi-mately 135 years old. In addition, the width of rings can indicate the occurrence of poor or good growth conditions. As shown in Figures 18 and 19, a good year produces a wide ring and poor years narrow ones. But ring

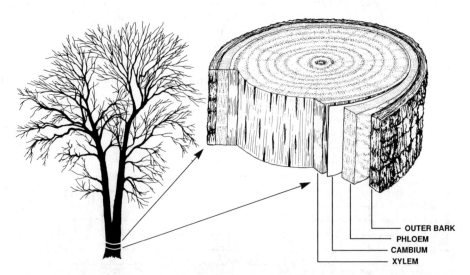

OUTER BARK
PHLOEM
CAMBIUM
XYLEM

Figure 16. Schematic illustration of a section of a tree trunk showing annual growth rings. It is the thin cambium layer that is the actual growth layer responsible for adding to the size of a tree.

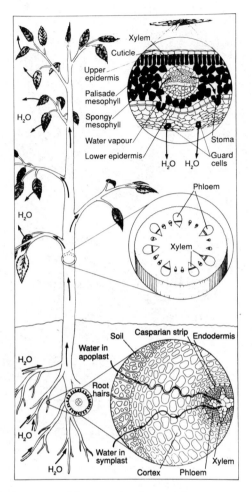

Figure 17. Simplified anatomical diagram of a plant and the pathway taken by moisture from soil to root hairs to xylem. (Courtesy of IPC Magazines, London, England.)

number and size are effective only with the life span of that tree. In order to extend dates further back in time, a continuum linking successive generations of trees was needed. Cross-dating was the answer. As shown in Figure 20, cross-dating is based on the fact that all trees in a specific region would have encountered the same or similar climatic stresses and would therefore contain similar patterns of wide and narrow rings, reflecting the yearly conditions of that region. To match these patterns across generations, thin cores are removed from a number of trees at a specific site and analyzed in the manner shown in Figure 20.

The bristlecone pine has far more than 10 rings. This species of conifer remains the oldest living form of life on earth. Several in California's White Mountains are known to be 4000–4500 years old. They were seedlings

Figure 18. Decades of good and poor years are revealed by the rings in this cross section of Douglas fir from the Southwest. (Courtesy of the Laboratory of Tree-Ring Research, University of Arizona, Tucson.)

when the Great Pyramids of Egypt were being built. And the titanic conifers, the giant sequoias, have been verified as being over 3500 years old and still growing. By using cross-dated wood fragments unearthed at archeological sites, chronological dating of areas has been extended back beyond the ages of the bristlecones, but not without their help, over 8000 years. The essential point is that dendrochronological dating is accurate to the year. And recently, tree-ring data have been used to ascertain past climates.

Using computer-assisted analyses, tree-ring widths have been converted into estimates of atmospheric pressure, temperature, and precipitation for years predating the existence of climate records. The result has been the development of a series of climate maps of the past. Studies at the Laboratory of Tree-Ring Research, University of Arizona, have led to several germane findings. Researchers there have divided past winters into five types, based on pressure patterns. Two of the five bring cold temperatures to the Northeast. Between 1900 and 1970, these two cold types occurred 30% of the time. For 300 years before, they occurred over 50% of the time. Over the entire 370 years, the two types could be expected to occur almost every second year. And over the same period the expectation of warm winters is better than half the cold. Clearly, variability is the natural rule, but it is exceedingly important to obtain the long view.

Additionally, drought patterns over the past 250 years have been revealing for the frequency of their occurrence. Perhaps most important for our purpose is the coincidence of periods of drought with alternate periods of minimum sunspot activity. They occurred at roughly 22-year intervals, during every other minimum in the 11-year sunspot cycle. Thus tree-ring data show that the most severe periods of drought coincide with or lag slightly behind alternate 11-year minimums of solar activity.

In addition to their rings, trees of course are a source of carbon, but carbon exists in the form of three isotopes: ^{12}C, ^{13}C, and ^{14}C, of which ^{12}C (the most abundant) and ^{13}C are the most common and stable forms. Carbon-14, the unstable isotope and least abundant, is known to decay away to nitrogen, and the rate of its decay is immutable. Thus a given

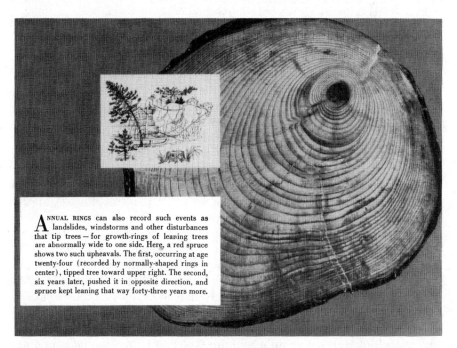

ANNUAL RINGS can also record such events as landslides, windstorms and other disturbances that tip trees — for growth-rings of leaning trees are abnormally wide to one side. Here, a red spruce shows two such upheavals. The first, occurring at age twenty-four (recorded by normally-shaped rings in center), tipped tree toward upper right. The second, six years later, pushed it in opposite direction, and spruce kept leaning that way forty-three years more.

Figure 19. Annual growth rings can tell a story. Some of the events in the life of the tree, whose rings are shown here, are noted. (Courtesy of the American Museum of Natural History, New York.)

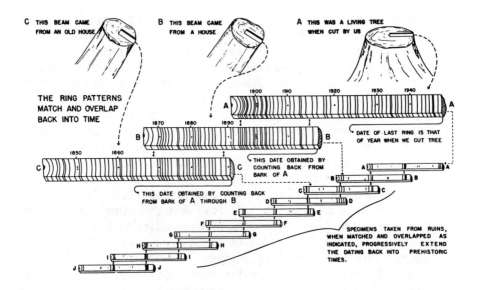

The principle of crossdating.

Figure 20. The principle of cross-dating, an ingenious procedure, is depicted here. (Courtesy of the Laboratory of Tree-Ring Research, University of Arizona, Tucson.)

quantity or unit decays to its daughter product in a known period. This regular and fixed process of decay is the basis of radiocarbon dating.

Radiocarbon, ^{14}C, is produced at the top of the atmosphere by neutron bombardment of atmospheric nitrogen atoms. ^{14}C atoms are rapidly oxidized to $^{14}CO_2$, which diffueses downward and mixes with the rest of the atmospheric CO_2 and eventually enters all vegetation via the photosynthetic process. Plants, and animals as well, assimilate a proportion of ^{14}C into their tissues, which is in equilibrium with that of the atmosphere, as there is a constant exchange of new CO_2 as old cells die and are replaced. When a tree or any living thing dies, this exchange and replacement ceases. From death on, the ^{14}C content of the tree declines as the ^{14}C decays away to nitrogen. In fact, with the death of the tree a radioactive "clock" begins ticking to extinction. It is a clock built to run down. The decline of ^{14}C follows a negative exponential rate in which 50% of the carbon-14 disintegrates approximately every 5730 years—the half-life of carbon. In 5730 years half of the ^{14}C decays away. Only half is left. With the passage of another 5730 years another 50% decays, leaving a quarter of the original amount, and on and on. Thus 10 half-lives equal the passage of 57,300 years, at which time the sample contains 0.001% of the carbon it contained before death. As the decay of carbon produces beta particles, which can be counted in an appropriate device, their number can be translated into a date when the sample was alive. And the ^{14}C content of tree rings may hold clues to past solar activity.

In the late 1950s Hessel L. DeVries, a Dutch dendrologist, reported a striking increase in the ^{14}C content of tree rings from the 1650s to the 1730s—the period of decreased to absent sunspot activity. As noted earlier, when solar activity is high, the earth's magnetic field shields the earth from the solar winds and their highly charged particles. Therefore fewer are available to slam into the upper atmosphere, creating neutrons that transmute nitrogen atoms to radioactive carbon. Thus an increase in ^{14}C indicates decreased solar and magnetic activity—a key concept. This finding has since been confirmed from tree rings across the earth, which adds credence to the Maunder minimum.

Using the bristlecone pine to obtain a chronology of ^{14}C fluctuations back to 5000 B.C., John Eddy was able to construct a history of solar activity. As he remarked, "every decrease in solar activity such as the Maunder sunspot minimum matches a time of glacial retreat." And he finds that "these results . . . make it appear that changes on the sun are the dominant agent of climatic changes lasting between fifty and several hundred years." He has yet another pregnant thought. "It would seem," he wrote, "that Maunder and Spörer were right and that most of the rest of us have been wrong. As is often the case in the onrush of modern science, we had too quickly forgotten the past. . . . Long ago we held that the sun was perfect, and when the telescope showed that it had spots on it, we took comfort in the thought that it was at least regular in its behavior. It now appears to be neither of these, and it is probably inconstant as well." The Maunder minimum and the Little Ice Age are more than abstract climatological phenomena. They do in fact bear witness to events of great human consequence.

It didn't take a great stretch of imagination for Erik (the Red) Thorvaldson to call the fertile coastal region where he and his band of Icelanders made their landfall, Greenland. They had sailed west from Iceland in A.D. 983, when the North Atlantic was warming and the pack ice disappearing. This was the beginning of the Medieval Maximum, a warm period that lasted for some 500 years and then gave way to the Maunder minimum and the Little Ice Age. During those five centuries, two Norse colonies flourished on the Greenland coast: the eastern settlement with a population of some 5000 at Greenlands's southern tip, and the western settlement with 1500–2000 at Gardar near Julianehaab. It was during this warm period that Erik's son Leif was able to lead a party of Vikings to the mainland of North America.

By the 15th century, temperatures had dropped again, approximately 2°C; the ice built up and sailing became difficult, if not impossible. Both colonies began to lose touch with their home country, and by 1500 the last of the colonists had perished. It need not have ended that way. For 50 years they were in close, often intimate, contact with the Eskimo, the Inuit people who had learned to thrive under the unsparing harsh conditions of the frozen north. But the Norsemen never adapted to the changing conditions. They persisted in cattle farming rather than tending sheep and goats or shifting to fishing, and they persisted in their European style of dress rather than adapting the warmer clothing of the Inuits. Flexibility and resilience were not part of their lifestyle, and they paid a bitter price for it. Theirs may be a cautionary tale for our time.

Climate change has indeed played a central role in human history, but whether the sun is at the center of the drama remains uncertain. Nevertheless, the evidence that the earth responds to solar activity is reasonably good, albeit circumstantial. Substantiation via more direct evidence will be required before modelers accept changes in their equations.

The quest to find links between the solar cycle and terrestrial climate persists, and the question stands. Could the sun be less than constant and predictable? Karen Labitzke of Berlin's Free University may have found a convincing link and brought us closer to an answer. It has long been known that stratospheric winds shift abruptly from a westerly phase to an easterly one and back again every 2–2.5 years. This shifting is called the quasi-biennial oscillation—QBO. A quasi-biennial oscillation refers to meteorological events that occur in cycles of 2–3 years; hence the modifier quasi for biennial. When the QBO is in its west phase, the whirling stratospheric winds that swirl over the North Pole in winter are stronger and colder than winds in their east phase. In 1987, Labitzke reported after a reexamination of past weather records that these atmospheric winds at the equator shift direction every 12–15 months, blowing first from the west, then reversing and blowing from the east. She found that when these winds are blowing from the west and sunspot activity is at or near its 11-year peak, the solar maximum, the southwestern United States has a colder than normal winter. That is, during the west phase the pattern was reversed; the swirling winds over the pole were weaker and warmer. In addition, during a solar minimum, the east phase was also reversed, becoming colder and stronger. Reviewing ground-level weather records, she also found that the effects noted at the polar region could be found in weather changes at sea level. Thus for the United States it meant that during a westerly phase of the QBO, the Southwest would be colder in winter—and she predicted that the winter of 1988–1989 would be colder in the Southwest. But January, February, and March proved to be warmer than usual. She attributed this to the effect of the strong El Niño currents, which overwhelmed the effects of sunspot activity. On the other hand, the occurrence of winter warming in both the United States and western Europe has shown remarkably strong statistical correlation with the solar cycle over the past 40 years when the reversal of wind direction is accounted for. In fact, she was alone in predicting the warming that made for the unusually mild winter of 1988–1989 in England and Europe. If a straightforward mechanism could be developed to account for a connection between solar and climatic variability, it would represent a formidable advance, a turning point that could not readily be ignored (29). More recently, Labitzke, together with Harry Van Loon of NCAR, demonstrated by Monte Carlo techniques that her original correlations were significant at the 95% confidence level or better. Unfortunately, the data cover only 36 years—3.5 solar cycles—a relatively short period where reliability is concerned. Should the correlation be extended over additional earlier cycles, it would gain wider acceptability (30).

The question of constancy has been approached from yet another direction. Most recently, Sallie Baliunas and Robert Jastrow of Harvard and Dartmouth, respectively, developed a 20-year record of 74 solar-type observations of the magnetic behavior of stars and suggested that Maunder minimum may be mirrored in the brightness variations of sunlike stars. In fact, brightness changes of more than 0.1% seem widespread. Of the 74 solar-type stars on which they have gathered data, 13 were looked at monthly since 1966, and nightly since 1980. The remaining 61 have supplied additional data since 1978. They maintain that "the reduction of sunspots observed in the 17th century may be a general characteristic of solar-type stars that is best described by a low mean level of magnetic activity, rather than the near disappearance of sunspots. The varying level of magnetic activity is a more quantitative indicator of changes in the sun's physical state than the historical record of sunspots" (31). And rather than continue to use the term Maunder minimum to refer to the specific observation of solar activity in the 17th century, they broaden the term to include solar-type stars exhibiting low levels of magnetic activity. Solar-type stars are stars similar to the sun in our solar system. These are referred to as main-sequence stars with masses and ages similar to our sun's. Alpha Centauri is one, Epsilon Eridani another, and Procyon A yet a third. In doing this, they found that 4 of the 13 stars appeared to be in a Maunder minimum at any one time, which was in good agreement with the carbon-14 record of solar magnetic activity, "which suggests that the sun has spent about one-third of its time in the past several millenia in magnetic minima." Furthermore, they now believe that there have been incorrect interpretations of the Maunder minimum. Their analysis indicates that "the level of magnetic activity is much lower in a Maunder minimum than in a sunspot minimum," and they conclude that when "a solar-type star enters or leaves a Maunder minimum, its brightness may change by more than 0.1 percent observed during the sun's 11-year cycle. ". . . A change," they tell us, "of only 0.22–0.55 percent has been estimated to be sufficient to account for the 0.4–0.6°C global mean temperature change in the Little Ice Age. Our results, which suggest the possibility of changes of several tenths of a percent, may have significant implications for climate change over century-long time scales." Should follow-up studies by others as well as Baliunas and Jastrow support this conclusion, it must seriously call into question the assumption of a constant sun in all calculations using GCMs for climate forecasting.

Moving beyond telescopic observations, Richard Willson of the Jet Propulsion Laboratory, California Institute of Technology, and his co-investigators at the University of California monitored solar irradiance aboard the Solar Maximum Mission satellite, using high-precision active cavity radiometers. They found irradiance to have been highly variable over a 5-month period. In fact, they reported two large decreases of up to 0.2% lasting about a week, and that these decreases were correlated to the development of clusters of sunspots. In addition to the two major dips, their instruments also recorded continuous variability (32). Among the goals of the Solar Maximum Mission is the obtaining of regular irradiance data over a 22-year period, one solar magnetic cycle, while another seeks to obtain highly precise short-term data—over minutes to months. Palpably, this is the type of data that will refine model projections as well as indicate whether the estimated 0.5°C mean surface temperature increase since 1860 is possibly related to variable solar output. Final data are expected by the year 2002.

As a group, modelers do not hold with the idea of a variable sun, but if recent findings are correct—that is, if well designed follow-up studies support the Labitzke and Baliunas–Jastrow data—then the 0.3–0.5°C warming that has been calculated to have already occurred can be explained as natural variability on a century time scale. And if human activity cannot overwhelm natural climatic phenomena, then the earth should be heading into a cooling period—the next glacial period. That is, if sunspot activity is directly related to global climate changes, then another lull is to be expected in the 21st century—and a lull would mean a cooling trend and another Little Ice Age. Clearly, the few tenths of a degree variation, which may be attributed to solar variability, begs for confirmation. Given the importance of the above observations, further research along these lines must be encouraged. The case for constancy remains unproved, but the vigor of research indicates the need to keep the question open, which must imply the persistence of uncertainty in current model forecasts.

ADDITIONAL APPREHENSIONS

It is also the fact of the spirited defense of the predicted numbers and their interpretation which gives pause. Are model projections entitled to overly strong defense? Climate modelers themselves have had reservations about their models. It is only with close scrutiny that models change and improve. Thus in 1976, Wang and co-workers of the NASA/Goddard Institute for Space Studies commented that "climate modeling is at a primitive stage and is not yet capable of reproducing interannual and long term climate variations. The primary difficulty is the large number of physical processes that come into play for time scales longer than the radiative time constant of the atmosphere, which is of the order of one month. These processes involving the atmosphere, ocean, cryosphere and land surfaces are particularly complex because of the significant interactions and feedback effects that occur among them over climate time scales (33).

Although James Hansen personally "went public" before a Senate committee in May 1988 with his strong statement about global warming, which swiftly became worldwide news and has since made greenhouse warming a household expression, he and his group of atmospheric physicists had published a forceful paper in the summer of 1981 predicting essentially the same consequences of burning fossil fuel. At that time they stated that "anthropogenic carbon dioxide warming should emerge from the noise level of natural climate variability by the end of the century." They went on to say that "the warming projected for the next century is of almost unprecedented magnitude. On the basis of our model calculations we estimate it to be about 2.5°C. . . . It would approach the warmth of the Mesozoic, the age of dinosaurs." Nevertheless, they were also aware of the coarseness of models and commented that "many caveats must accompany the projected climate effects. First, the increase of atmospheric CO_2 depends on the assumed energy growth rate, the proportion of energy derived from fossil fuels and the assumption that about 50 percent of anthropogenic CO_2 emissions will remain airborne," and finally that "the predicted global warming for a given CO_2 increase is based on rudimentary abilities to model a complex climate system with many non-linear processes" (34). And again in 1988, Hansen and his colleagues at GISS were confident in their model's prediction that a global temperature increase beyond 0.4°C would occur during the early years of the 1990s. If that occurred, they were confident that it would represent "convincing evidence of a cause–effect relationship." But they did not leave it there. They went on to call for more intensive research. "Major improvements," they wrote, "are needed in our understanding of the climate system and our ability to predict climate change. We conclude that there is an urgent need for global measurements in order to improve knowledge of climate forcing mechanisms and climate feedback processes. The expected climate changes in the 1990s present at once a great scientific opportunity because they will provide a chance to discriminate among alternative model representations, and a great scientific challenge because of demands that will be generated for improved climate assessment and prediction" (4). Too often these "caveats" are given short shrift rather than the attention they deserve. Actually, they should receive coverage equal to that of the predictions. Without taking cognizance of them, unwarranted conclusions are too often drawn. Caveats ought to be included in abstracts and conclusions of published papers; otherwise cursory readings—the "quick study"—will too often lead to false impressions and assumptions of more than is in fact there. With the coming of the mid-1980s, the number of model studies had proliferated enormously. It was time for a peer review and evaluation of the state of the subject.

In 1986, a group of scientists at the Lawrence Livermore National Laboratory's Atmospheric and Geosciences Division reported the results of their detailed, in-depth analysis of published studies of climate change. Among their conclusions they argue persuasively that "this analysis . . . has demonstrated that climate is both variable and complex and that the data for documenting past climate changes have many shortcomings." They went on to pose the relevant question of the accumulated forecasts. "Has the planetary surface," they asked, "actually warmed?" They went on to answer it. "We conclude that the global mean surface temperature has indeed been increasing, albeit erratically, since the nadir of the Little Ice Age and certainly over the period of recorded data . . . and we have confidence in this conclusion." They continued: "The fragmentary early data suggest significant cooling prior to 1883 such that 25–50 percent of the subsequent warming may represent a return to earlier levels. . . . There remains a strong inclination to assign the current warming to increased CO_2 because of the simultaneity of the two occurrences and because of the inability to identify any other warming mechanism." The fact of the Little Ice Age's continuance so close to our contemporary period is often overlooked or forgotten. That the calculated 0.5°C warming over the past 100 years may actually reflect nothing more than a return to normal, deserves greater attention. Their following observation is singularly trenchant. "Pending supporting information this inclination should be resisted for two reasons," and they went on to specify that ancient climatic data as well as the medieval climatic optimum had attained temperatures above those currently in place without increases in CO_2, and further that increased CO_2 is not the only anthropogenic process that could effect a planetary warming. They were, they stated, "left with the clearly unsatisfactory conclusion that while we are witnessing a warming of the terrestrial climate, we cannot now identify its cause. Since higher temperatures appear to have prevailed earlier through natural processes, this current warming too may be natural. However, even if it is of anthropogenic origin, it need not be due only to increased CO_2. But if it is due predominantly to CO_2, then our present climate models require work to reconcile them with observational data on patterns of surface temperature change" (35). It is unfortunate that reviews such as this in highly respected refereed journals rarely attain wide public dissemination. The information languishes, to the detriment of public discussion and decisionmaking. Of course, reviews may not delight those whose published work comes in for further scrutiny. But it must be done if the subject is to advance and mature.

Recently, Kiril Kondratiev of the Laboratory for Remote Sensing, Institute for Lake Research, Leningrad, commenting on the Ellsaesser/Laurence Livermore Group assessment, concluded that because there was no guarantee that the current estimated warming could not be ascribed to a return to a normal, milder climate, and that evidence for a release of 100–200 gigatons of carbon as a result of forest cutting and agriculture took place prior to 1938, whereas CO_2 additions due to fossil fuel combustion (about 175 gigatons of carbon) occurred mainly after 1938, "there are no grounds for persuasively ascribing the climate warming to the impact of an increased CO_2 concentration."

Following the detailed review of the Lawrence Livermore Group, Schlesinger of the Department of Atmospheric Sciences and Climate Research Institute, Oregon State University, and John F. B. Mitchell of the United Kingdom's Meteorological Office performed a sweeping review of energy balance and radioactive convective models as well as general circulation models in terms of their equilibrium climatic response to increased carbon dioxide. Their assessment is not only germane and trenchant but raises a critical question. Listen to them. "How," they ask, "can we be confident in the GCM projections of CO_2-induced equilibrium climate change?" Their response is carefully reasoned. "To have confidence . . . requires that these models correctly simulate at least one known equilibrium climate, with the present climate being the best choice because of the quantity, quality and global distribution of contemporary instrumental observations." "However," they continued, "an evaluation of the fidelity of a GCM in simulating the present climate is not simple for a variety of reasons, including how well the simulated and observed climate represent their corresponding equilibrium climates and the poor quality of the observations of many climatic qualities such as precipitation over the ocean and soil moisture. However, forgetting the difficulties for the moment, suppose that a GCM simulates the present climate perfectly. How can we gain confidence in its ability to simulate another climate different from that of the present? This of course is the conundrum—the current dilemma. In the case of weather forecasting, this question can and has been answered by making thousands of forecasts and comparing them with the actual evaluation of the weather. Unfortunately, this cannot be done for climate because only a few paleoclimatic reconstructions have been made, and these may or may not be of sufficient quality to provide a meaningful assessment of the GCM's capability. Thus there is an inherent limitation in our ability to validate the accuracy of GCM perturbation simulations, which thereby affects our confidence in the accuracy of the GCM simulations of CO_2-induced climate change. The state of the are is that GCM's simulate the present climate imperfectly, although some of the models do reasonably well . . . yet these models frequently employ treatments of dubious merit, including prescribing the oceanic heat flux, ignoring the oceanic heat flux, and using incorrect values of the solar constant. Such approximations indicate that the models are physically incomplete and/or have errors in the included physics. Furthermore, the state of the art is that the CO_2-induced climate changes simulated by different GCM's show many quantitative and even qualitative differences; thus we know that not all of these simulations can be correct, and perhaps all could be wrong" (36). Coming as this does from acknowledged leaders in the field, such an assessment must be taken seriously. It cannot be wished away.

Perhaps it is time to consider simplifying models. Are they inherently overly complex, and does this provide yet an additional source of uncertainty? Some few years ago O'Neill suggested a relationship for thinking about model complexity. Given the current heavy reliance on models for decisionmaking, it may be appropriate to resurrect his model. In Figure 21, we see that as a model increases in complexity, the α error or systematic bias declines, and as additional parameters are included, β error or measurement error can be expected to increase. The resultant model inaccuracy γ ($= \alpha - \beta$) may tend to increase as model complexity increases for exceptionally complex models. Thus if the net result of the two sources of error is scrutinized, it is entirely possible that an optimum level of model complexity will be found for minimizing prediction error. This could be a model approach to modeling (37).

Earlier the question was raised of whether more or less is to be expected of climate models vis-à-vis other models. A realistic appraisal seems to be that given their greater complexity, their relative immaturity, and the demands being placed on them, it does appear that while much can be expected of them, at this stage of their development more is being asked than can reliably be delivered. A case in point: from their model runs Schneider and Chen of NCAR (38) predicted in 1980 that sea levels would rise from 4.6 to 7.6 meters (15–25 feet), and along with this rise there would be property damage in excess of $1 trillion (1970 dollars) from coastal flooding in the continental United States alone. In 1990 dollars the losses would be greater by a factor of 100 or 1000. That prediction obtained wide currency. By 1989 however, Schneider was of the opinion that sea level would rise by no more than 0.5–1.5 meters (1.6–4.9 feet). Most

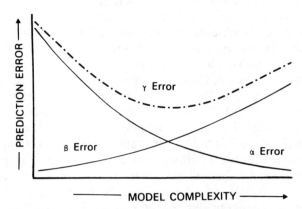

Figure 21. This plot of the relationship of prediction error versus model complexity suggests that one can anticipate increased error as complexity increases. On the other hand, it also suggests that an optimum level of complexity can be obtained (35).

recently, estimates held that between 1990 and the year 2030 mean sea level could be expected to rise by 1.5–2 feet: 18–24 inches over 40 years. These are quite different estimates and would perforce elicit different responses. Scientifically this is as it should be: evolving change with increasing knowledge. The books are never closed or completed on any subject but are always added to and redefined as new data become available and are reviewed and evaluated. Thus the change in understanding of sea level height vis-à-vis global warming and subsequent loss of glacial ice to meltwater, and the dissemination of this information, are highly appropriate. Sea level rise is one of the most severe of the predicted dislocations of a warmer earth and as such ought to be enveloped in the best available data. What is inappropriate is the misuse of premature data, especially for forecasting. Science and scientists are not and have never been prophets. Foretelling the future is for astrologers and palm and tea-leaf readers, not scientists. And climate models were not meant to be predictive tools. They were developed to learn about the processes that generate climate. They do not incorporate crystal balls, tarot cards, or necromantic mumbo jumbo. Had national or international policy decisions been based on the earlier prediction of sea level rise, would a major error have been made? Is the current prediction of up to 24 inches subject ot change over the coming years? Of course it is. It could be revised up or down as new insights on melting of ice sheets and thermal expansion of the oceans become available. Of this we can be sure—data will become firmer and thus more suitable for planning purposes. A rush to judgment must be resisted.

At the conclusion of the Workshop on Greenhouse-Gas-Induced Climate Change: A Critical Appraisal of Simulations and Observations, held in Amherst in May 1989, the participants issued a press release, which stated in part that "progress to improve our ability to project future climate will best be achieved by the further development, analysis and verification of climate models, by the acquisition, assembly and analysis of climate data, by observational studies of climate processes, and by providing the human and computer resources required for these tasks. We . . . conclude that the need to reduce the current uncertainties about the magnitude, timing and regional detail of future climate changes is an urgent international priority."Clearly, there are groups of scientists who are satisfied with the predictions their models produce, and others who are not. For anyone with even a nodding acquaintance with the experimental domain, this is the normal state of affairs and is to be desired for its constant state of agitation, which motivates further work to resolve differences. Problems occur when for whatever reason interested parties pick and choose conclusions that suit their current predilections—a practice that defeats the purposes of science.

Most dictionaries define the adjective "uncertain" and the noun "uncertainty" as unable to be accurately known or predicted; not to be depended on or unreliable. This would appear to be a fair estimate of the current state of the science. It certainly does not mean or imply that with the passage of time uncertainty will not give way to certainty, as assumptions are replaced by vigorous data; it will. But not for at least 10 years—not a great deal of time considering the system and the magnitude of the problem. But it does mean that uncertainty as now understood should not be brushed aside as of little consequence; that but for a few bits of relatively inconsequential data the issue would be brought into sharp focus for the more finicky among us.

It probably would be well to consider a recent remark by Francis Bretherton, Director of the Space Science and Engineering Center, University of Wisconsin. Bretherton was quoted as saying that "at our present level of knowledge, experimentation is more important than realism." This has the ring of understanding and reality to it. It testifies to both the state and needs of the science. There is no doubt that knowledge of the climate system is growing rapidly and that quite literally the best minds in the country and around the world have "turned to" this demanding issue, but it is also true that firm statements about future climate are premature—even tentative. What then are policymakers and the public to do—how are they to respond when faced with scenarios, such as we now have, in which general circulation models running with instructions to double atmospheric greenhouse gas concentrations respond differently? Of three highly sophisticated models—Oregon State University (OSU), NOAA/GFDL, and NASA/Goddard—two, NASA/Goddard and OSU, project large excess rainfall patterns in summer in the Southeastern states, while NOAA/GFDL predicts a decrease. Two of the three, NASA/Goddard and NOAA/GFDL, show drought conditions in Kansas, Iowa, and Nebraska, while OSU forecasts an increase in precipitation. Which is correct? Are any? The OSU model predicts drought conditions in southern California; the others forecast adequate or increased levels of rainfall. Is there any way to choose among them? Hardly. Unless the uncertainty inherent in each model is reduced quickly, and together the three groups are able to determine where the differences between them lie, which should be given the highest priority, the margin for error, for everyone, is increased Less Time is then available for corrective measures; further exacerbating a discomforting situation. Unless and until these differences are resolved, governmental policymakers and corporate leaders can continue to affirm a "business as usual" attitude. How can they not? It would be expecting too much for them to adopt a "damn uncertainty, decisions now" attitude. On the other hand, advocates of decisive action now accept model forecasts as they are. Are we thus hoist on the horns of a dilemma? Perhaps. Perhaps not, if we consider Reid Bryson's admonition. Bryson, senior scientist at the Center for Climatic Research, University of Wisconsin, considering the current state of models and their predicting ability, offers what I believe is an appropriate cautionary tale for forecasts of climate change for the mid-21st century. "Let us first note," he said, "that a statement of what the climate is going to be in the year A.D. 2050 is a 62-year forecast. Do the models which are used as a basis for the forecast have a demonstrated capability of making a 63-year forecast? No. A 6.3 year forecast? No. A 0.63 year forecast? No. Have they successfully simulated the climatic variation of the past century—and a half? No. Is the handling of the radiative transfer in the model tested against reality? No. Do the models take into account aerosols in the atmosphere? Rarely. They are marvels of mathematics and computer science, but rather crude imitations of reality. . . . I do not say that the GCM-type models are useless or wrong, when they are the most sophisticated known way of considering the bulk of atmospheric physics simultaneously over the whole interacting globe. However, this does not mean that they are complete or entirely correct. . . . There is simply much more which needs to be done before they are sufficiently reliable for climate forecasting" (39).

This sentiment has surrounded model forecasting almost from its inception. For example, in a seminal paper on the question of an increase of atmospheric CO_2, Revelle and Suess remarked that "so little is known about the thermodynamics of the atmosphere that it is not certain whether or how a change in infra-red back radiation from the upper atmosphere would affect the temperature near the surface" (40). Although this was written more than 30 years ago, it appears no less true today. Presumably, impetus for gathering in-depth data on levels of CO_2 proceeds from the Revelle–Suess admonition that "human beings are now carrying out a large scale geophysical experiment of a kind that could not have happened in the past nor be reproduced in the future." They continued with the statement that "within a few centuries we are returning to the atmosphere and oceans the concentrated organic carbon stored in sedimentary rocks over hundreds of millions of years." That was surely a key. But the crux of the current concern was also contained in their report. "This experiment," they iterated, "if adequately documented may yield a far-reaching insight into the processes determining weather and climate. It therefore becomes of prime importance to attempt to determine the way in which carbon dioxide is partitioned between the atmosphere, the oceans, the biosphere and the lithosphere." This of course was a basis for much of the research that occupied part of the International Geophysical Year of 1957–1958, including Keeling's establishing the CO_2 monitoring station atop Mauna Loa. But it was their conclusion, following an extensive mathematical treatment of the rate of exchange and absorption of CO_2 between sea and atmosphere, which must be reread and reconsidered. For them it was clear that in the coming decades "a total increase of 20 to 40 percent in atmospheric CO_2 must be anticipated. This should certainly be adequate to allow a determination of the effects if any of changes in atmospheric carbon dioxide on weather and climate throughout the earth." That was three decades ago; ample time, if today's prognostications are no less meaningful, to have elicited some climate signal. Their "if any" creates further consternation. Apparently, the great lack of knowledge about the climate system allowed for a lack of or no untoward response. That is worth a second thought, for if it is so, it should be of continuing concern in that current models backed by supercomputers continue the idea of a climate response in the future—by the third, fourth, or fifth decades of the 21st century. But they have gone a step beyond. Today there is no "if any." Today there is certainty. By 2030 the temperature *will* increase by 1–2°C with dire consequences for humankind. Perhaps. And by the final decades of that century, average temperature may well rise as much as 5°C—a level of warmth not seen on earth since the hegemony of the great dinosaurs. Again, perhaps. Were the Revelle–Suess calculations off by several decades, or were the exceptionally warm 1980s the signal?

That was in 1957. In 1982, writing in a more popular vein about carbon dioxide and world climate, Revelle again reminded and cautioned us that "about the only facts available are the actual measurements of atmospheric carbon dioxide . . . [but that] steps should be taken to obtain more evidence and to consider the consequences of a continuing increase in atmospheric carbon dioxide" (41). At about the same time that Revelle offered this observation of the state of model forecasting, the Executive Committee of the World Meteorological Organization meeting in Geneva ventured their evaluation of then current predictions: "The increasing amount of CO_2 released into the atmosphere as a result of human activity may have far reaching consequences on the global climate, but the present state of knowledge does not permit any reliable prediction to be made of future CO_2 concentrations or their impact on climate."

More recently, Eric J. Barron, director of the Earth System Science Center at Pennsylvania State University and editor of *Global and Planetary Change*, commented that "the single greatest obstacle to predicting in detail the weather of the future is the complexity of the climate. Even the most sophisticated computer models fall far short of fully representing the elements—oceans, land-masses, vegetation, pollutants—that determine weather patterns at a particular time and place." "What's more," he tells us, "it is difficult to distinguish, on the basis of models alone, between man-made trends in global climate and manifestation of large-scale natural cycles, about which little is known. And neither the mathematical equations that underline the models nor the available meteorological data can be used to project with certainty how global warming, once it sets in, will alter other aspects of the climate" (42).

This uncertainty theme pervades the scientific literature on climate change, but one would never know it for all the impact it has had on current discussions or public perception. What is so remarkable is the lack of attention it receives.

A decade ago, Robert W. Kates of Clark University, who would probably describe himself as a social scientist, was invited to attend the World Climate Conference, sponsored by the World Meteorological Organization.

His remarks at the conference echo throughout the years. Time has not dulled their pungency; vexing questions seem to have a life of their own. Particularly prescient for us are portions of his concluding remarks: "Finally," he told the assembly, "I wish to dispose of a certain fallacy before we enshrine it as a myth. Implicitly, and occasionally explicity, the suggestion has been made here that the end result of a world climate program would be to have such scientific understanding of climatic change, variation, and impact that rational, objective decisions can be made." And he continued, saying, "we actually know a great deal about decision-making and what we know does not suggest that this simple view, often labelled as the 'economic' or expected utility model, will prevail. The great choices that our emerging climatic understanding will pose will always be made under conditions of uncertainty. They will be made in the face of conflicting information by nations and individuals with conflicting goals. In the face of such uncertainty, choices that depend on climate will be evaluative rather than cognitive and we would be well advised to consider such value judgments directly." As he concluded he asked the conference to consider that "currently we have no completely rational or objective way to foresee the future, yet for a week we have discussed the climatic impacts of the year 2000, 2050 or even 2400. Proper economic analysis always discounts the future against the present and even at the smallest reasonable discount rate the future quickly becomes valueless. But unfortunately we do not behave as if we believed in such analysis. The concern for future generations (as well as for ancestors) is a part of our common heritage. How to assert this value, how to recognize it in a rational calculus that denies it, is a troubling and vexing issue. There is a healthy dialectic between fact and value. It is to be hoped that fact will inform our values and narrow the dilemma of our choices" (43).

Both the level of emission of the trace IR absorbers as well as their rate of removal remain uncertain. Consequently, predictions of anticipated temperature increases by the 22nd century, the year 2100, depend on guesstimates. Thus, for example, Figure 7 offers a set of projections based on low, medium, and high rates of emission. As the figure shows, the period 1860 to 1986 (dark black dots) is reasonably well established. Each of the projected trends from about the year 2000 to the end of the century (2099) is shown. The meaning is inescapable; no matter what the level, a rise is foregone, and the timing is emission dependent. Jaeger tells us that the consensus of considered opinion holds that there is a 90% chance that the actual value would fall somewhere between the high and low level (44). But that is pure intuition on their part. Nevertheless, what is suggested is that two of the three estimates are decidedly bad news. Three-tenths of a degree per decade would result in a 1°C rise by about 2022, 2°C by 2060, and 3°C by 2099. For a 0.8°C decadal increment, a rise of 1°C would occur by the year 2000, 3°C by 2020, and 5°C by 2040. Although this may be glossed over as no more than an intellectual exercise, it does contain the seeds of discontent. At no time since the last interglacial period some 150,000 years ago, that is, beyond the last glacial period, has the earth experienced more than a 2°C increase. We've had no experience with such large rises. We've experienced lower temperatures—as much as 5°C during the height of the last glacial period—but never increases. The fact of an increased warm period over the coming decades is already being referred to as a super interglacial age. Thus Figure 7 holds out potentially difficult times for those living beyond the year 2000; for by then, if the most severe of the scenarios proves to be the correct one, a 1°C rise in mean surface temperature will have occurred—with yet higher temperatures in the offing. Considering that 2000 is but 9 years hence, the possibility is irritating. Indeed, Figure 7 is irritating.

Data such as these must be a nightmare for policy people. Since it is well nigh impossible to prepare for the range of possibilities, policy for a 0.06°C decadal increase would be quite different from that required for an 0.8°C increase. Just how useful can such planning be? It would not be unrealistic to imagine a set of responses on the part of a policymaker—anger, despair, frustration, paralysis. Unless complete faith is placed in such projections, the reluctance to take precipitous action is understandable. Given that hope springs eternal, and that the lower scenario—0.06°C increase per decade —could be endured, it would probably be the one prepared for. More than likely it would not be a "business as usual" scenario. Adjustments in fossil fuel use would be required.

The possibility of a super interglacial period warrants further comment. Figure 22 describes the temperature of the earth for 150,000 years before the present and takes it another 25,000 years into the future. The cycles of cooler and warmer periods are evident. Natural variability is the rule. Had carbon dioxide and other IR absorbing gases not become excessive, the world, according to the best available evidence, should be near or at the decline of the current interglacial period, and over the next 15,000–20,000 years should slide into the next glacial period. This suggests that over the coming centuries humankind should begin to experience a cooling trend before the onset of a deep freeze some 20,000 years hence. That is what the natural climatic schedule calls for if the reason for seasons is correct. What Milankovitch could not foresee was interdiction by the hand of man. Possibly.

If excessive CO_2 enters the atmosphere and traps IR energy, heating the troposphere, the current interglacial as shown in the figure will develop a blip—what Wallace Broecker, professor of Geochemistry at Columbia University's Lamont–Doherty Geological Observatory, dubbed a "super-interglacial." this aberration should last 1000–2000 years, the time required for the carbon cycle to cleanse the atmosphere of the excess carbon dioxide. During this period, Broecker believes that the global mean temperature would rise several degrees higher than experienced on earth over the previous million years. After this warm interlude, the cooling trend

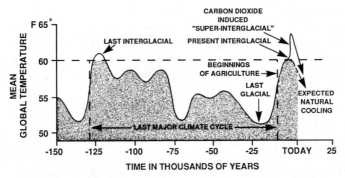

Figure 22. The time line of global mean surface temperature is shown for the past 150,000 years. Shown too is a potential CO_2-induced "blip" of several thousand years prior to the onset of the next glacial period.

would proceed as previously scheduled. If it is to be ice, obviously there's lots of time to prepare for it. If it's heat, time is short—that is, if this scenario is reality based.

In a state-of-the-art review of the global carbon cycle published in 1990, a group of scientists at the Oak Ridge National Laboratory complained that they were "unable to balance all the fluxes of the global carbon cycle over the period of 1800 to the present, and different mathematical models give results that are hard to reconcile." They believe too that "the research of the past few years has uncovered more complexities than were previously appreciated." Herein may well reside the difficulties with model projections; knowledge of the climate system is severely limited. A great deal of additional information is required to enhance forecasting ability. For example, the Oak Ridge group specifically identifies several major shortcomings. "For the purposes of understanding the carbon cycle and predicting future atmospheric levels of CO_2 . . . it is essential that we understand how terrestrial vegetation and ocean processes respond to changes in CO_2 and climate . . . and additional data and more comprehensive models are needed to link the climatic effects of the atmosphere–ocean system to geochemical events." Thus their conclusion is readily understandable. For them, "the dynamic responses of natural systems to CO_2 remain a puzzle—and the earth's climate may hang in the balance" (45). Some people would describe the current state of model forecasting as playing with less than a full deck. It is unfortunate that the same elements continue to be a source of frustration. Perhaps this chorus of concern needs buttressing, of a kind only an internationally prominent organization can supply, to obtain the attention it warrants. And perhaps the Intergovernmental Panel on Climate Change (IPCC) is such an organization. In the conclusion and summary to its report, *Climate Change: The IPCC Scientific Assessment*, the IPCC is forthright about the needs and has this to say about models: "There is clear need for further improvement of the accuracy of climate models. . . . Much further experience needs to be gained in the design of coupled models in order to avoid the equally unsatisfactory choices of accepting a progressive climate drift or of empirically correcting the behavior of the upper ocean. . . . The validation of a number of atmospheric model variables has been handicapped by limitations in the available observed and model data. . . . Future assessments would benefit from improved estimates of precipitation and evaporation over the oceans, and of evaporation, soil moisture and snow depth over land. . . . The lack of appropriate data has also severely hindered the validation of ocean models, [and] adequate data on the seasonal distribution of ocean currents and their variability, and on salinity and sea ice thickness are especially needed." Finally, they remarked on the need for the type of data which would "contribute to our understanding of how to distinguish between natural climate fluctuations and changes caused by increased greenhouse gases" (46). The gaps are evident, the shopping list is long, but above all there is awareness and openness about inherent limitations.

Without a doubt a considerable amount of excellent work has been accomplished by modelers. Great skill and intellectual prowess have been demonstrated in efforts to represent global climate. They deserve full praise for their often heroic efforts. But with the "cascade of uncertainty" that current models possess, they remain an unfinished symphony. As a class, all general circulation models are possessed of a similar affliction. Their construction, from the ground up, as it were, is almost entirely dependent on atmospheric, physical, and chemical data. Given their extent and tight coupling, the degree of incorporation of biologic and oceanic data is far too limited. It may be that incorporation of such data would be no more than fine tuning, which would not be inappropriate, but it may also be that their incorporation at a more sophisticated level would alter interpretations and predictions. Obtaining this type of information, as demonstrated by the recent observations of Ramanathan and Collins, should be a top priority for modeling research.

Uncertainties in current models notwithstanding, the IPCC report maintains that a global warming will occur by the middle of the 21st century, and that it will increase global mean surface temperature from 1.5 to 4.5°C. It also suggests that over the last century there has been a warming of from 0.3 to 0.6°C. Their best guess is that by 2030–2040 the warming will be closer to 2.5°C than 4.5°C. It is well to bear this 2.5°C in mind because small mean temperature changes can carry an explosive charge, as it were. If past history is a guide and if the paleoclimatic records have been correctly interpreted, the difference in temperature between the last glacial periods and the current interglacial was only 5°C. So seemingly small temperature fluctuations can induce extreme temperature effects. Last time it was cold and ice. This time, an unscheduled warming trend may be on the way toward interdicting the scheduled cooling cycle. Since there is no historical precedent for this, the potential and possible social, economic, and ecologic dislocations can only be surmised. But if the warming effects bear any relationship to the cooling, the changes could be prodigious.

A major concern of a warmer earth is its predicted rate of occurrence. Changes of global temperature of 1–5°C are, as we have seen, not new. They have occurred many times in the geologic past. The difference today is that rather than doing so over 50,000, 100,000, or more years, time enough for plant and animal communities to adapt and evolve, current projections are for a warming within 50–100 years; hardly enough for adaptation and/or migration.

As yet, greenhouse-induced warming has not been observed. Clear signals have not been heard or seen. Nevertheless, the potential threat for inhabitants of planet Earth is unlike any threat faced at any time since humankind turned from hunting and gathering to agriculture. Perhaps that is why so little concern appears to be given to it. It may simply be incomprehensible—unlike a dam bursting, the nuclear bombing of Hiroshima and Nagasaki, the eruption of Mount St. Helens, the destruction of Pan Am Flight 103, the sinking of the *Titanic*, World War II and the Holocaust, crime in the streets, or air pollution. These are all mentally comprehensible, manageable events with which people can come to grips. Even here there is a spectrum from relatively mild to severe, and even here such things as the Holocaust remain for some unbelievable and incomprehensible. But excessive heating within the greenhouse, the atmosphere, with consequent increased global temperature of only 1, 2, or 3 degrees, is meaningless for the majority of the world population. For most people an increase of 1–4°C might seem mild, even inviting. Even the underlying implications of additional very hot days and drought are not all that objectionable. People do understand that life goes on, and on, and that what is needed are adjustments, and that these will be provided. They always have. And the notion that somehow driving cars and trucks, burning coal and oil, and clearing land could raise the temperature of the earth enough to make life unbearably warm—warm enough to melt the snow and ice of the polar regions and thereby raise sea levels sufficiently to make coastal cities of inland cities—is no more than the stuff of science fiction. The whole idea lacks any semblance of credibility.

For James Hansen, director of the Goddard Institute for Space Sciences, the frustration must be palpable. He and others have produced data they believe appropriately forecasts the changes in climate the world can anticipate by the years 2040 and beyond. He is confident his data are correct but he has no way of proving it—especially to cautious world leaders. The impasse seems intolerable and is in no one's best interests. In "their heart of hearts" many scientists must realize that they are asking the world to take a leap of faith which reality-based people are reluctant to take, having never before confronted a similar situation. The result could be paralysis. It needn't be. But what if . . . what if over the coming decades atmospheric

CO_2 levels increase from their current 350 ppm level to 600 or 700. And what if methane and the CFCs increase along with it; what then? What are the problems that could ensue? Or perhaps more to the point, what are the dislocations that have been projected, and does a high carbon dioxide atmosphere necessarily bode ill for mankind? Are there no advantages?

REFERENCES

1. Rind, D. Rosenzweig, A., and Rosenzweig, C. Modelling the Future: A Joint Venture. *Nature* 334:483–486, 1988.

2. Menabe, S. and Wetherald, R. T. Thermal Equilibrium of the Atmosphere with a Given Distribution of Relative Humidity. *J. Atmos. Sci.* 24:241–259, 1967.

3. Manabe, S., and Wetherald, R. T. The Effects of Doubling the CO_2 Concentration on the Climate of a General Circulation Model. *J. Atmos. Sci.* 32:3–15, 1975.

4. Hansen, J. E., Fung, I., Lacis, A., Rind, D., Lebedeff, S., Ruedy, R., and Russell, G. Global Climate Changes as Forecast by Goddard Institute for Space Studies Three Dimensional Model. *J. Geophys. Res.* 93 (D8):9341–9364, 1988.

5. Hansen, J. E., and Lebedeff, S. Global Surface Air Temperatures: Update through 1987. *Geophys. Res. Lett* 15:323–326, 1988.

6. Kuo, C., Lindberg, C., and Thompson, D. J. Coherence Established Between Atmospheric Carbon Dioxide and Global Temperature. *Nature* 343:709–913, 1990.

7. Bradley, R. S., and Jones, P. D. Data Bases for Isolating the Effects of the Increasing Carbon Dioxide Concentration. In: M. C. Mac Cracken and F. M. Luther (Eds.). *Detecting the Climatic Effects of Increasing Carbon Dioxide.* U.S. Department of Energy, DOE/ER-0235. NTIS, Springfield, VA, Dec. 1985.

8. Quadfasel, D., Sy, A., Wells, D., and Tunik, A. Warming in the Arctic. *Nature* 350:385, 1991.

9. Mercer, J. H. West Antarctic Ice Sheet and CO_2 Greenhouse Effect: A Threat of Disaster. *Nature* 271:321–325, 1978.

10. Zwally, H. J. Breakup of Antarctic Ice. *Nature* 350:274, 1991.

11. Doake, C. S. M., and Vaughn, D. G. Rapid Disintegration of the World Ice Shelf in Response of Atmospheric Warming. *Nature* 350:328–330.

12. Cess, R. D., Potter, G. L., Blanchett, J. P., Boer, G. J., Ghan, S. J., Kiehl, J. T., Letreat, H., Li, Z. X., Liang, X. Z., Mitchell, J. F. B., Morcrette, J. J., Randall, D. A., Riches, M. R., Roeckner, E., Schlese, U., Slingo, A., Taylor, K. E., Wetherald, R. T., and Yagai, I. Interpretation of Cloud–Climate Feedback as Produced by 14 Atmospheric General Circulation Models. *Science* 245:513–516, 1989.

13. Slingo, A. Wetter Clouds Dampen Global Greenhouse Warming. *Nature* 341:104, 1989.

14. Mitchell, J. F. B., Senior, C. A., and Tugram, W. J. CO_2 and Climate: A Missing Feedback? *Nature* 341:132–134, 1989.

15. Ramanathan V., and Collins, W. Thermodynamic regulation of ocean warming of cirrus clouds deduced from observations of the 1987 El Niño. *Nature* 351:27–32, 1991.

16. *Scientific Perspectives on the Greenhouse Problem.* George C. Marshall Institute, Washington, DC, 1989.

17. Barnett, T. P., and Schlesinger, M. E. Detecting Changes in Global Climate Induced Greenhouse Gases. *J. Geophys. Res.* 92:772–780, 1987.

18. Munk, W. H., and Forbes, A. M. G. Global Ocean Warmings: An Acoustic Measure? *J. Phys. Oceanogr.* 19:1765–1778, 1989.

19. Karl, T. R., and Jones, P. D. Urban Bias in Area-Averaged Surface Air Temperature Trends. *Bull. Am. Meteorol. Soc.* 70:265–270, 1989.

20. Quinn, W. H., and Neal, V. T. El Niño Occurrences Over the Past Four and a Half Centuries. *J. Geophys. Res.* 92 (C13):14.449–14.461, 1987.

21. Zheng, D. W., Song, G. X., and Luo, S. F. El Niño Prediction. *Nature* 348:119, 1990.

22. Smith, W. O., Jr., Codispoti, L. A., Nelson, D. M., Manley, T., Buskey, E. J. Niebauer, H. J., and Cota, G. F. Importance of Phaeocystis Blooms in the High-latitude Ocean Carbon Cycle. *Nature* 352:514–516, 1991.

23. Kirchman, D. L., Suzuki, Y., Garside, C., and Ducklow, H. W. High Turnover Rates of Dissolved Organic Carbon during a Spring Phytoplankton Bloom. *Nature* 352:612–614, 1991.

24. Tans, P. P., Fung, I. Y., and Takahashi, T. Observational Constraints on the Global Atmospheric CO_2 Budget. *Science* 247:1431–1438, 1990.

25. Detweiler, R. P., and Hall, C. A. S. Tropical Forests and the Global Carbon Cycle. *Science* 239:42–47, 1988.

26. Wigley, T. M. L. Could Reducing Fossil-Fuel Emissions Cause Global Warming. *Nature* 349:503–506, 1991.

27. Latham, J., and Smith, M. H. Effect on Global Warming of Wind-Dependent Aerosol Generation at the Ocean Surface. *Nature* 347:372–373, 1990.

28. Eddy, J. A. The Case of the Missing Sun Spots. *Sci. Am.* 236:80–89, 1977.

29. Labitzke, K. Sunspots, the QBO, and the Stratospheric Temperature in the North Polar Region. *Geophys. Res. Lett.* 14:535–537, 1987.

30. Labitzke, K., and Van Loon, H. Association Between the 11-year Solar Cycle, the QBO, and the Atmosphere. Part II. Surface and 700 mb in the Northern Hemisphere in Winter. *J. Climate* 1:905–920, 1988.

31. Baliunas, S., and Jastrow, R. Evidence for Long-Term Brightness Changes of Solar-type Stars. *Nature.* 348:520–522, 1990.

32. Willson, R. E., Gulkis, S., Janssen, M., Hudson, H. S., and Chapman, G. A. Observation of Solar Irradiance Variability. *Science* 211:700–702, 1981.

33. Wang, W. C., Yung, Y. L., Lacis, A. A., Mo., T., and Hansen, J. E. Greenhouse Effects Due to Man-Made Perturbations of Trace Gases. *Science* 194:685–689, 1976.

34. Hansen, J. E., Johnson, D., Lacis, A., Lebedeff, S., Lee, P., Rind, D., and Russell, G. Climate Impact of Increasing Atmospheric Carbon Dioxide. *Science* 213:957–966, 1981.

35. Ellsaesser, H. W., MacCracken, M. C., Walton, J. J., and Grotch, S. L. Global Climatic Trends as Revealed by the Recorded Data. *Rev. Geophys.* 24:745–792, 1986.

36. Schlesinger M. E., and Mitchell, J. F. B. Climatic Model Simulations of the Equilibrium Climatic Response to Increased Carbon Dioxide. *Rev. Geophys.* 25:760–798, 1987.

37. O'Neill, R. V. Error Analysis of Ecological Models. In: D. J. Nelson (Ed.). *Radionuclides in Ecosystems* (USAEC–CONF 710501), pp. 898–908. NTIS, Springfield, VA, 19XX.

38. Schneider, S. H., and Chen, R. S. Carbon Dioxide Warming and Coastline Flooding: Physical Factors and Climatic Impact. *Annu. Rev. Energy* 5:107–140, 1980.

39. Bryson, R. A. Civilization and Rapid Climatic Change. *Environ. Conservat.* 15:7–15, 1988.

40. Revell, R., and Suess, H. E. Carbon Dioxide Exchange Between Atmosphere and Ocean and the Question of an Increase of Atmospheric CO_2 During the Past Decades. *Tellus* 9:18–27, 1957.

41. Revelle, R. Carbon Dioxide and World Climate. *Sci. Am.* 247:35–43, 1982.

42. Barron, E. J. Earth's Shrouded Future. *The Sciences* 20:14–20, 1989.

43. Kates, R. W. Climate and Society: Lessons from Recent Events. *Weather* 35:17–25, 1980.

44. Jaeger, J. Developing Policies for Responding to Climatic Change. WMO/TD No. 225, April 1988. (A Summary of the Discussions and Recommendations of the Workshops held in Villach and Belagro, 28 Sept.–13 Nov. 1987.)

45. Post, W. M., Peng, T. H., Emmanuel, W. R., King, A. W., Dale, V. H., and DeAngelis, D. L. The Global Carbon Cycle. *Am. Sci.* 78:310–326, 1990.

46. *Climate Change: The IPCC Scientific Assessment.* WMO/UNEP IPCC. J. T. Houghton, G. T. Jenkins, and J. J. Ephraums (Eds.). Cambridge University Press, Cambridge, 1990.

CHAPTER 4

POTENTIAL DISLOCATIONS AND ADVANTAGES

Half a century ago, 100 million acres of the Great Plains dried up and blew away. "Dust Bowl" was added to our language and we learned that nature was not all that benign. Farmers in Oklahoma, Texas, Kansas, Colorado, New Mexico, and Nebraska wondered aloud about what might be "if it rains." For many people, that remains more than a memory. For others, it was too long ago to be of concern. The climate changes predicted to occur by the mid-21st century as a consequence of uninterrupted use of fossil fuel are expected to shift global mean air temperature higher than it has been at any time during our long, relatively warm period—the past 12,000 years, even beyond that to 120,000 years ago and the last interglacial period when the temperature was 1–1.5° higher than it is today. The higher temperature and increased atmospheric CO_2 levels of the next century are expected to affect plants, animals, and people—all living things will adapt or die. But this is not a new scenario for planet Earth. Climate shifts, as noted in Chapter 1, have occurred often and with regularity and have induced cyclic modifications in both terrestrial and marine environments such that time and again the existing flora and fauna have had to migrate to more hospitable climates. Climatic stability has not been the rule. Variability is nature's rule. Consequently, it is more than likely that many existing species evolved under environmental conditions different from today's and thus have developed the type of genetic constitution capable of adjusting to shifting ecologic stresses. But changes in climate have directed not only changes in flora and fauna as well as their ranges, but extinctions as well.

Sixty-five million years ago, half of all life on earth became extinct. Dinosaurs vanished along with 90% of all protozoans and algae; but mammalian ancestors of human beings survived. In the fossil record, this mass extinction marks the boundary* between the Cretaceous and Tertiary pe-

---

*It is called K-T rather than C-T to avoid confusion between Cretaceous and Carboniferous, which is abbreviated as C. K is from the German *Kreide* for Cretaceous.

riods of paleohistory—the end of the Mesozoic era. Paleontologists use the term K-T boundary to refer to this period. If this boundary is in fact the result of a cataclysmic event such as a giant meteorite hurtling through space and smashing into the earth with a speed of 360 miles per minute, there should be some record of it. An event so destructive of living things must leave some trace.

In 1979, Walter Alvarez, a geologist at the University of California (Berkeley), removed samples of sedimentary rock from the Appenine Mountains in the area of Gubbio, north of Perugia, Italy. These samples were of the K-T boundary in which a thin layer of clay separated limestone from the Cretaceous and Tertiary periods. On chemical analysis, the clay was found to contain large amounts of iridium, a mineral second only to gold in density and the most corrosion-resistant metal known—rare on earth but highly concentrated in meteorites. The level of iridium in the clay was as much as 160 times greater than in the limestone above and below it. That much iridium cannot be expected to arise from the surface of the earth. This suggested that a meteorite had in fact crashed into a warm, tropiclike earth, creating dense clouds of dust and debris, along with fire and smoke to a degree sufficient to block incoming sunlight—which lowered the temperature. Mass extinctions of dinosaurs and marine animals and plants occurred as a consequence of loss of photosynthetic activity and abrupt temperature decline.

Meteorite extinction is a theory supported by existing bits and pieces of reasonable evidence, but it is vigorously opposed by scientists who point quite rightly to volcanic eruption as a rich source of iridium; and that volcanic ash, dust, and gas can also darken the sky, reducing both temperature and photosynthesis. It is noteworthy that, as recently as 1785, Benjamin Franklin commented on the hard winter and cool summer of 1783, with its months of continued cloudiness, and such demonstrated lack of sun that "when its rays were collected in the focus of a burning glass [a magnifying lens], they would scarcely kindle brown paper" (1). Franklin wondered if the vast quantity of smoke continuing to issue from the volcanic eruption in Iceland was not the problem.

Was it volcanic action or a crashing extraterrestrial object that caused the mass dying? The distinction is more than academic. Understanding the actual cause of the extinctions could suggest the cause of other extinctions that have occurred, as well as the potential effects that milder climatic shifts may exert on living things. This is the type of information that could suggest effects of temperature increases induced by excessive trapping of heat by infrared absorbing gases. Bear in mind that if it turns out to be volcanic, in which case it would also mean high amounts of CO_2 released with consequent increase in temperature, it would suggest that the deaths were the result of heat rather than cold. To this end, two groups of scientists, one from the University of California who supports the meteorite extinction theory and a group from Dartmouth College who champion the volcanic theory, recently gathered samples in Gubbio in hopes of reaching agreement on the distribution of iridium. Each group's samples were coded and shipped to neutral laboratories for analysis.

But the excitement generated by the Gubbio trails may well have been eclipsed by more recent findings that further strengthen the impact theory. The newest and as yet strongest evidence of a collision between a massive comet or asteroid and Earth some 65 million years ago comes from tektites, tiny glass fragments recently discovered in Haiti. These fragments are free of gas and water, which would preclude volcanic origin because volcanic eruptions are gas driven. Their chemical composition, 60–67% silicon dioxide(SiO_2) also denies volcanic origin. Additionally, these fragments show unique fractures or "shock" lines in their crystal structure. This shocked condition appears only to arise from great heat and pressure—that is, impact (2). Furthermore, the site of impact of an extraterrestrial body has come to light. The Gulf of Mexico and the Caribbean area are now the sites of intense scientific activity. Primacy is being given to the Yucatan Peninsula, where a crater, some 60–100 miles in diameter, lies at a depth of 1000 meters off the Merida/Progresso coast. Its impact could readily have scattered glass shards as far away as Cuba and Haiti and could have liberated 10^{15} moles (10^{16} grams) of CO_2 from vaporized marl sediments into the atmosphere—comparable to the annual level of CO_2 emitted currently.

A number of mechanisms can explain the sudden extinction. A fireball of incandescent gas created by the explosion would have propelled terrestrial debris well up into the troposphere. Impact-generated dust would have been so dense that darkness would have occurred and lasted for months—long enough to cause widespread death. Without sunlight, photosynthesis would stop and the food chain would collapse and, along with it, all animal life dependent on it. If the impact occurred in the ocean, huge amounts of water vapor would have been hurled into the atmosphere to remain there far longer than dust particles. As an infrared absorber, this added water vapor could easily have caused additional warming of the troposphere with concomitant temperature rises on land—to a level intolerable to many species, but especially the poikilothermic dinosaurs. Those species that survived the darkness/cooling/lack of food phase would have succumbed to intense warming. If the impact occurred in areas of large limestone formations, such as mail and coral reefs, then CO_2 would have been released in great quantity, also producing warming. The heating of the atmosphere by the shock waves of the initial impact would have been sufficient for the oxygen and nitrogen in the air to combine, forming nitric acid, a severe form of acid rain. This strong inorganic acid could well have been responsible for the extinction of many marine invertebrates whose calcium carbonate shells are readily acid soluble. Finally, the widespread existence of soot in K-T-boundary clay suggests a far flung fiery conflagration as a consequence of impact—perhaps half the world's forests ignited and burned—a veritable firestorm of global proportions. If this didn't kill the animals directly, the soot and smoke would add to the debris to further darken the heavens. Thus at least four reasonable mechanisms can account for the massive extinctions on land and sea.

Now paleontological field studies show clearly that, contrary to earlier claims, dinosaurs and other species were still thriving at the time of impact. Growing evidence points to the fact that older paleontological explorations were wrong in believing that dinosaurs died out gradually. This does not mean that dinosaurs would be alive today if not for the impact. It does mean that the impact theory has, and continues to gain, wide support among paleontologists. It is also clear that no species lives forever.

Meanwhile, research goes on. Patrick Brenchley and co-workers at the University of Liverpool recently identified closely spaced extinctions spread over a period of 500,000 years, which they believe represent different stages in the cooling and warming of the earth's climate. Their data suggest that the extinctions can be attributed to global change related to the ice ages and are unlikely to be the consequence of meteorite impact. But the Brenchley group has no data indicating why a mild climate reverted to a cold one. When the answers are in and one theory prevails over another, if so clear-cut an eventuality can be imagined, we shall know how climate changed naturally and be able to reinstruct model programs. The more is "known" about past climate, the better predictions of the future can be anticipated. The essential point is that climate varies naturally, has shifted frequently, often abruptly, and may do so again. This time, however, it may be anthropogenically induced, and the question uppermost in the mind of scientists of a variety of disciplines is what dislocations could be expected from a temperature increase of as much as 5°C—especially if it occurs abruptly—in less than 1000 years, more like 80–100 and perhaps less.

Although for 1990, atmospheric levels of CO_2 are poised at 350 ppm and rising, with parallel increases in methane, water vapor, the nitrogen oxides, and the CFCs, which exert far greater warming effects than CO_2, for all intents and purposes, the current equivalent or effective CO_2 level may well be 400 ppm or greater. Thus, since the preindustrial era, circa 1800, with a CO_2 level between 260 and 290 ppm depending on whose values are cited, the increase has been dramatic. If 290 ppm is an appropriate value, then the increase to date has been approximately 21%. If we use 260 ppm, the increase over the past 190 years has been closer to 35%. But if we use 190 ppm, the level attained during the last glacial period, then it is patent that we are currently more than halfway to a doubling of our atmospheric CO_2 level. Curiously enough, given model predictions, a concomitant increase in global temperature has not manifested itself for so large a CO_2 increase. It is quite conceivable that temperature lags behind increases in CO_2 and that the predicted increases will be observed within 10–15 years. It is also conceivable that the manifestation will occur in another guise.

What environmental or ecological changes can be expected to occur? Both model predictions and experimental studies suggest a variety of responses that could affect our health, food supply, and general existence. Nonetheless, predicting responses to a warming can be dicey, since the predicted temperature increases exceed that of anytime since the retreat of the great ice sheets. Furthermore, the variety of biotic responses to warming is both unlimited and, worse, unknown. Our archives lack any such experience. Too often, modelers assume predictable responses for ecosystems, which paleoecologists find disturbing, especially as ancient fossil records now reveal that past climate changes affected different species differently. Changes in rainfall, for example, that should accompany temperature increases are certain to affect plants, that is, crops and agriculture. But over the centuries, we have learned to expect year to year fluctuations, which suggest the wide-ranging effects that local weather can exert on farming. The Dust Bowl exemplifies this, but unseasonable frosts and drought occur with sufficient regularity to ruin crops and harvests and thereby adversely affect the food supply. Indeed, increasing ambient CO_2 levels will surely lead to increased per acre yields of a number of crops and the warmth and precipitation could easily affect growing seasons, which should have much to say about the general availability of food.

Before pursuing potential ecologic and human dislocations, it is worth considering the problem of temperature constraints for living things. Biological systems, human beings and other species, however high or low taxonomically, are inherently poised so that almost any change in the environment represents a challenge or threat to survival. Such things as temperature and humidity can affect the welfare and efficiency of living systems, by upsetting the stability of the "internal environment," which is the essential condition for living cells. In a word, homeostasis, the maintenance of metabolic equilibrium, can be fatally jeopardized.

The metabolic processes of cells are highly susceptible to changes in temperature. All living things dwell comfortably within a narrowly defined temperature range. Every species has an optimum temperature at which its metabolic pathways function most efficiently. If it rises to a certain point above that, chemical reactions can so speed up that the energy is consumed faster than it can be replenished. If temperature falls too low, chemical reactions slow down and simply stop. In both instances, death ensues. One of the reasons why mammals have been so environmentally successful is that their body temperature has been independent of external temperature. Unless living things can stabilize their own temperature, their cells will be forced to respond to changes in external temperature. So, for example, poikilotherms (from the Greek poikilos, meaning varied—as in varied temperatures), cold-blooded animals such as amphibians and reptiles, have systems in which internal temperature rises and falls with the external temperature. Mammals and birds, on the other hand, are said to be homoiothermic (from the Greek homoio, meaning like or same, as in "of like temperature") or warm-blooded, their body temperature being independent of their surroundings. When chilled, their blood vessels constrict, withdrawing warm blood from the cooler surfaces—in effect, automatically insulating itself to prevent heat loss. When body temperature begins to rise, blood vessels dilate, permitting the blood to lose its heat readily. Sweating is another mammalian physiological response for shedding heat and cooling the system. As Claude Bernard, professor of physiology at the College DeFrance, maintained years ago, "the constancy of the internal environment is the absolute condition for a free life." Plants are not unlike animals in physiological compensation, but there is a limit to adjustment. The fact of freezing temperatures, as well as heat and lack of moisture on crops, plants, and flowers, is lost on no one. An excellent example is the death of flowers and plants each fall with the first "unseasonable" drop in temperature. From our accumulated experience over the centuries, we have learned to expect fluctuations, not annually, but certainly with some regularity and frequently enough to have made lasting impressions—beginning with Joseph and his brothers in Biblical accounts. Here then the concern for significant increases in temperature takes on new meaning. Perhaps this is the message to be extracted from predictions of a 2 or 3°C or higher warming. The numbers are small, but so is the line between life and death.

Generally, models are consistent in predicting that the impacts of warming will be neither uniform around the world nor necessarily equally harmful. Canada could well profit, should drought once again stalk the Great Plains and the wheat- and corn-producing areas of the U.S. "bread basket" shift northward. Similarly, the Soviet Union, desperately trying to wrest minerals, timber, grain, and vegetable crops from frozen Siberia for a sagging economy, would welcome a thaw in that part of the world. Conversely, it would rejoice if predictions hold and the wheat-producing Ukraine dries along with the American heartland. The effect of such shifting is exemplified in the accompanying cartoon (Figure 1), which also suggests that opportunities and benefits may be the obverse side of the warming coin.

Thus modelers have projected that, at the high northern latitudes— above 65°N—warming will be 50–100% greater, especially during winter, and substantially less than the global mean in regions of sea ice in summer. Therefore, if predictions are correct, temperatures could rise as much as 18°F in the north polar area. Rainfall is projected to increase in winter by 5–10% on those continents situated at middle and high latitudes from 35 to 55°N. Consequently, the United Kingdom and northern Europe generally can be expected to become even wetter than they currently are. North Africa and southern California are expected to become drier. In fact, rainbelts may already be moving northward away from the more southerly areas such as the region stretching across those countries making up the

Figure 1. This cartoon says it all for the possibility of rising sea levels. (Reproduced with permission of the cartoonist A. J. Meder.)

Sahel: Senegal, Mali, Niger, Chad, the Sudan, Ethiopia, Somalia, and Djibouti. Rain has steadily decreased there for 30 years and appears to be heavier in the northern latitudes. Heating of the atmosphere changes the way the winds blow, which has a major effect on location of clouds and, with them, available moisture.

In assessing global climate responses to a warming trend, a panel of experts recently convened by the National Academy of Sciences offered the following appraisal: "Increased heating of the surface will lead to increased evaporation and, therefore, to greater global mean precipitation. Despite this increase in global average precipitation, some individual regions might well experience decreases in rainfall." And they went further, saying that "thermal expansion of seawater along with melting or calving of land ice should manifest itself in a rise of global mean sea level." It must, however, be borne in mind that much of the conjectures of potential ecological dislocations are truly intellectual exercises, which must be undertaken because CO_2 increases may well have effects far beyond temperature shifts; plants and animals may respond positively to a doubling and even tripling of atmospheric CO_2. But no one has had real-time experience with a warmer world, so that projections can reasonably be viewed as conjectures. We do know, however, that rapid climate change occurred at the end of the last glaciation. But then it warmed to approximately the present pleasantness and allowed humankind to mature and prosper. The question that remains is what changes and extinctions may be expected with a rise in temperature—a warming beyond anything previously known. Given that all* living things exist within a fairly narrow range of temperature, 0–40°C (32–104°F), higher temperatures could reasonably be expected to exact a toll, depending, of course, on how much higher. However, and perhaps unfortunately, ecologists have had little, if any, experience with large-scale environmental disruptions and certainly non involving abrupt climate change. For example, recent studies in the vicinity of Mount St. Helens' eruption show that recovery in the blighted areas is proceeding at a far greater pace than anticipated. Apparently, a variety of species have survived scorching and burial under a foot of ash. Forest service ecologists believe that classic or traditional ecological recovery studies focused on fields used for farming rather than natural landscapes, which appear to have faster recovery potential.

The problem that looms large for many biological scientists is that, in the past, when climate changed, as it did regularly and frequently, naturally, it did so over long time periods—tens of thousands to hundreds of thousands of years. The impact of an atmosphere enriched in CO_2 and other greenhouse gases may be to reduce contemporary climate change to years (decades)—too little time perhaps to permit plants and animals, including insects, the necessary time to adapt. Without adequate time, what biological and ecological dislocations could be expected of natural communities? That is but one of many nagging questions. For example, are there sufficient data and experience on which to draw appropriate inferences and, worse, conclusions about what to expect? In greenhouse warming scenarios, pundits have in effect been "fighting the last war." Predictions appear to flow from things as they are, without consideration being given to adaptation to change. Can it be expected that we human beings will sit still for global warming? Will we not respond with imaginative offsetting strategies—inventions and discoveries with which to make life not only possible but agreeable?

This chapter seeks to bring these seemingly disparate ideas together. While it is essential to consider potential impacts such as rising sea level, storms, and agricultural impacts generally, as well as effects on human health, how large an impact is anyone's guess. Given the possibility of dire consequences for mankind, it is unfortunate that "guess" remains the operative term.

If not the most widely feared, potential increases in the level of the seas may certainly be one of the most dislocating consequences of global warming. In terms of the planet, all land areas are islands—all are surrounded by water. Historically, it was the coastal areas that were the first to be populated and although migration inland followed, coastal areas around the world remain heavily populated and continue to be the venue of major

---

*Not all. Thermophilic bacteria are known to thrive in hot springs where temperatures of 45–96°C (113–204°F) occur.

industries. Significant increases in sea levels could inundate large areas, making them uninhabitable and forcing large population migrations, with their potential for both domestic and international turmoil, to deal with the uprooted. A rise in sea level of 3 feet—1 meter—could affect as many as 300,000,000 people, and the cost to prevent a dislocation of this magnitude could easily run as high as \$25–30 billion per year for the dikes and barriers needed to keep rising waters at bay. Unfortunately, estimates also indicate that half the countries at risk cannot afford to underwrite the costs of protective systems.

As shown in Figure 2, the United States, for example, has a long and potentially vulnerable coastline. But cost estimates to protect these areas run approximately \$150–200 million, which can readily be defrayed by the benefits of erecting protective barriers in these highly productive areas. On the other hand, costs to a country such as Bangladesh, where some 50–60 million people are at high risk, in the low-lying coastal areas, are estimated at \$500–700 million. But there is little economic productivity to offset these staggering costs. Given its unique topographical nature, in- cluding the large delta areas, a rise in sea level of 1 meter would flood fully 15% of the total land area.

In addition to sea walls and other types of barriers, coastal cities such as Charleston, South Carolina; Miami and Pensacola, Florida; Mobile, Alabama; and New Orleans, Louisiana need to consider drainage systems, early warning detection systems, emergency repair procedures, and regu- lar maintenance, as well as mechanisms for managing the movement of

Figure 2. Location of wetland and coastal sites along the U.S. border, which are potentially at risk of rising seas. These sites are the result of a regional/national analysis developed by a general simulation model. (Courtesy of the U.S. EPA, Washington, DC.)

large numbers of people, should a need for evacuation arise. However, experience with such concerns is not at ground zero. The Netherlands has been dealing effectively with hostile seas for hundreds of years and has prospered. It may be necessary for some coastal areas to adopt the practice of building homes on poles or pilings and to increase the use of a variety of boats and ships as has been the custom in Venice. The many and varied sophisticated technologies available—and to be created—notwithstanding, rising sea levels could easily mean the displacement of millions of people; inland within their own countries, or away to less vulnerable areas. The question that cannot now be answered is whether they will be accepted or turned away and whether horrendous problems will ensue. A rise in sea level of 50 centimeters, approximately 20 inches, less than 2 feet, could well inundate such low-lying areas as Canton, China, with its huge population, and the Maldives (a group of exceedingly flat islands in the Indian Ocean, southwest of India), as well as the Bangladesh delta, forcing their populations to seek living space elsewhere. The 200,000 people of the Maldives are all at risk. Because examples such as these are believed to be real possibilities, considerations are being put forward for gradual relocation of coastal populations to forestall such eventualities. But uprooting large populations requires time, well thought out educational programs, and money. All are often in short supply. The margin for error is small. If done badly, civil strife could ensue.

For Florida, a 20-inch rise could move the high water line inland 200–400 yards—up to one-quarter of a mile. But in Louisiana, the same 20-inch rise would move it inland as much as 2–3 miles. Thus low-lying coastal areas would be affected differently by a similar rise. And beach areas such as Ocean City, Maryland, would lose 100–200 feet, which doesn't seem like much, but its beach-front high-rise development over the past 20 years has already taken away much of the previously existing beach, leaving only a relatively narrow ribbon of sand. Thus a 20-inch rise in ocean level would completely submerge the existing beach and bring the water to the front door of many of the apartment buildings.

In Chapter 3, it was noted that early estimates of sea level rise were some 5–7 meters. It was those estimates that touched off much of the concern for rising waters and their draconian effects. Since then, new estimates have reduced the expected rises to one-third of the original predictions. But the margin for error remains large, and thus uncertainty stalks all approaches to planning. This is the reality and it is what must be lived with, given the lack of real data.

Sea level, like climate, varies naturally. During glacial periods, ten of them over the past million years, sea level has been lowered by as much as 500 feet as tremendous volumes of water were drawn up and frozen into the great ice sheets. During the last interglacial period, 100,000 years ago, sea level appears to have been about 6–7 meters (18–20 feet) higher than current levels. Melting of the great ice sheets and freezing their immense content of water may or may not occur but concern is real enough. The amount of water locked away is truly enormous. An example may provide an idea of the dimensions involved. In September 1987, an iceberg currently designated B-9 broke away from the Ross Ice Shelf in the vicinity of the Bay of Whales, in Antarctica. When first spotted by NOAA's *Landsat* satellite, it was 96 miles long, 22 miles wide, and 750 feet deep. It was as large as Long Island . . . and floating around. According to scientists at Columbia University's Lamont–Doherty Geological Observatory, it contained 287 cubic miles of fresh water—enough to provide the entire population of the world with two 8-ounce glasses of water for the next 2000 years! And this is a relatively small iceberg. The amount of water locked away in ice borders on the incomprehensible. Thus lack of concern for melting must be seen as an aberration. Currently, however, the ice that covers Antarctica contains about 90% of the world's ice. Its melting would surely alter both sea level and climate. But although projections of sea level rise include meltwater from ice sheets and glaciers, present estimates indicate that the Antarctic ice sheet is stable—in equilibrium. As glaciers and ice sheets dump ice into the sea, snow falling over Antarctica replenishes it at about the same rate. Therefore, for some time to come, sea level rise may be limited to thermal expansion.

On the other hand, sea level may not only *not* rise, but it may fall with atmospheric warming. Researchers at the University of Colorado and University of Wisconsin believe that the great ice sheets should increase in volume as a consequence of water from the seas being locked into the ice. This seeming contradiction would occur as warmer air holding more moisture loses its moisture as it falls as snow over the colder polar regions. This increase in snowfall would account for a lowering of sea level of from 0.1 to 0.3 meters by the end of the 21st century—A.D. 2100. Bentley and Giovinetto of the University of Wisconsin's Geophysical and Polar Research Center have calculated present sea level change as dropping 0.3–0.8 millimeters per year, accumulating a total drop of from 0.1 to 0.3 meters by the end of that century (3).

In a related study, David A. Robinson of Rutgers University and Kenneth F. Dewey of the University of Nebraska recently found snow cover to be shrinking over the northern hemisphere. During 1988 and 1989, it was at its lowest extent since reliable snow cover monitoring by satellite began in 1972. Snow is a pivotal variable in the global climate system, influencing the global heat budget through its effect of increasing surface albedo reflectivity. The changes observed seem to imply that the winter snow season begins a little later and ends a bit earlier in the northern United States and Canada—but not so much that anyone would actually notice (4). Whether the observed decrease is due to a CO_2-forced warming is still too early to tell, but their findings do coincide with a decade of warmer temperatures. Since satellite-gathered data provide the type of comprehensive view that ground-based temperature and snow measuring stations cannot attain, the

Robinson–Dewey measurements could provide a critical sentinel indicator of future climate change. They also point out that warmer temperatures would lead to extensive low-pressure areas, which also imply increased snowfall in the more temperate latitudes. Again, the increased moisture in the atmosphere will have had to come from the seas, which would suggest a lowering of sea level as snow piles up and freezes into ice. Thus, at this juncture, the fact of a warmer climate and accompanying increases in sea level, is not only not a foregone conclusion, but a drop in sea level cannot be ruled out. At this juncture, uncertainty muddies the crystal ball. Nevertheless, one thing is reasonably certain; a sea level rise of 3 feet (1 meter) is not anticipated by the mid-21st century. However, even a modest rise of as much as 6 inches, due primarily to thermal expansion, would initiate a panoply of problems in low-lying coastal areas around the world, should protective measures fail to be in place.

EFFECT OF RISING SEA LEVEL ON WETLANDS

Wetlands are yet another concern of rising sea levels. Rising seas could inundate wetlands, permitting penetration of salt water further inland as well as increasing coastal erosion. Louisiana, with 40% of all U.S. wetlands, is a case in point. Its wetlands, the area between yearly high tide and mean sea level, are largely disappearing. Each year, as much as 40 square miles sink. The loss is formidable. It is the wetlands, the marshes, that support the country's largest fin and shellfish industry, producing some 30% of the total harvest. Fish landed in the Gulf of Mexico spend the early part of their lives in Louisiana marshes. And these marshes are the wintering ground for most of the migratory waterfowl that travel the Mississippi flyway.

In addition to rising sea level, loss of marshes or wetlands comes from subsidence of bottom land, which was formed from sediment washed into the delta areas by the Mississippi River. So much silt has accumulated that it sinks due to its own weight. In some areas, as much as 1 inch per year submerges. Thirty years ago, subsidence was a cyclical process; the weight of the sediment forced the ground to sink, but the river continued to add new sediment each spring, resulting in a round of sinking and refilling. Since the 1950s, dams across the many tributaries have held back the sediment so that currently only half the cycle functions—that is, sinking. Add to this the eroding effects of tropical storms and hurricanes and the devouring of roots of protective, binding marsh plants by the Coypu, or Nutria, a beaverlike rodent imported from South America to control water hyacinths clogging canals, and there is in place a formula for the loss of wetlands. With the loss of wetlands to the open sea, salt water intrusion also kills the roots of plants holding the marsh together. Thus, before any contribution that global warming might make to higher sea levels and to increased winds and storms—all deleterious to wetlands—wetlands are already in imminent danger. Global warming would only further exacerbate the problem.

STORM AND HURRICANE ACTIVITY

Earlier, it was noted that floods and storms are instantaneous events that are all but impossible to predict and difficult to study and analyze. Floods are the result of the combination of torrential rains and the physical characteristics of the drainage area through which rushing water must pass, as well as the changes made in these natural areas by unfettered urban-style development. Given the pressure to use all available land, including known and established flood plains, floods can be expected to exact an increasing toll on life and property. Although the decade of the 1980s had been a relatively quiescent one in terms of such losses, predictions of global warming for the 1990s and beyond include increased frequency of severe storms and hurricanes.

Hurricanes are, in fact, tropical cyclones that originate over tropical oceans and that blow in a large spiral around a relatively calm center or "eye." Thus hurricanes can be likened to giant whirlwinds in which air moves in an immense, compressing spiral around a center of inordinate low pressure, attaining maximum velocity in a band extending as much as 30–40 miles from the rim of the eye. It is not unusual for such a storm to prevail over tens of thousands of square miles. In the northern hemisphere, the cyclonic circulation takes the form of counterclockwise rotating wind with an intensity of at least 74 miles per hour. The area of strong winds assumes a circular to oval shape, often as much as 500 miles in diameter. Hurricanes are characterized by inundating rains, as much as 10 inches falling in an area during its swift passage. Considering that 1 inch of rainfall over an area is equivalent to over 27,000 gallons of water, the downpour is quite literally overwhelming. The severity of a hurricane is classified on the Saffir–Simpson Hurrican Intensity Scale into categories 1–5. As Table 4.1 indicates, category 3 hurricanes have peak winds up to 130 miles per hour (mph); category 4 up to 155 mph, and category 5 exceed 155 mph. Categories 3–5 are considered major hurricanes. For the first 78 years of the 20th century, over 50 category 3–5 hurricanes stormed U.S. coastal areas. And during that same period, another 75 category 1 and 2 storms struck the coasts. Over the past 10 years, the numbers have declined sharply. Nevertheless, when one such as Hugo barrels into a coastal city as it did to Charleston in September 1989, the devastation can be appalling. Damage ran to hundreds of millions of dollars; thousands were driven from their homes and hundreds lost their homes. When Hurricane Gilbert, the greatest storm ever to strike the western hemisphere in the 20th century, swept over the Cayman Islands on September 13, 1988, its winds exceeded

Table 4.1
Saffir–Simpson (S-S) Hurricane Intensity Categories

S-S Category	Maximum Sustained Wind Speed (miles per hour)
1	74–95
	(33–42) [a]
2	96–110
	(43–49)
3[b]	111–130
	(50–58)
4	131–155
	(59–69)
5	155+
	(769)

[a] Meters per second in parenthesis (1 meter/second equals 2 miles/hour).
[b] Category 3, 4, and 5 storms are 9–25 times more destructive than category 1 and 2 storms.

130 mph. As it raced southwest toward Mexico's Yucatan Peninsula, it gathered speed and force. Striking the provence of Quintana Roo and its principal city, Cancún, 24 hours later, it had attained category 5 status but seemed bent on establishing a new category as its wind attained 179 mph. And it wrought damage in direct proportion to its unimaginable power. In one area, a 75-ton fishing trawler was lifted out of the water and deposited close to a nearby hotel, which was itself demolished. In the wake of its passage over and through Jamaica, it not only left 500,000 people homeless but totally destroyed the poultry and banana industries.

Until 1988, Camille held the record for the United States. It was a category 4/5 hurricane, which was directly responsible for 256 deaths as it roared across Louisiana and Mississippi in 1969. Hurricanes named after Hura Can, the Cavil's Indian God of Evil, are the earth's most violent storms, and they have a legacy to prove it.

Over the decade of the 1980s, southeast coastal areas from North Carolina to Texas have experienced large population growth. People from across the United States have moved there in such large numbers that they have taken on new importance. Curiously enough, in a country where risk to life has become all but unacceptable, hurricane damage potential is growing as coastal cities increase in population. As shown in Table 4.2, between 1980 and 1990, the increases have been noteworthy: from +0.8% in Louisiana to +33.0% in Florida. Without concomitant protection against the ravages of hurricanes and flooding, both people and property are fair game for losses, injury, and death. Even though there is persuasive indication that hurricanes could well increase in number and intensity, in fact, they are happily a rare natural phenomenon. They remain one of science's more perplexing and yet unsolved mysteries (5). They are unique to planet Earth. Nothing out of this world compares with them or, more to the point, contains the essential ingredients required for their development.

Recently, the hurricane model developed by Kerry Emmanuel of the Center for Meteorology and Physical Oceanography, MIT, revealed that, with increased temperature, the intensity of hurricanes could be expected to increase (6). And digging through historical records, J. E. Coleman appears to have found that during periods of above-average sea-surface temperatures, the frequency of storms increased in the North Atlantic. While these are suggestive of future changes, vis-à-vis global warming, there is a need for caution. Increased storm intensity and frequency may well occur, but they cannot properly be laid to climate change. Recall the certainty with which the withering summer of 1988 was heralded as a warming signal when, in fact, it was the consequence of an El Niño event.

In a recent accounting of hurricane activity, William H. Gray, of the Department of Atmospheric Sciences, Colorado State University, found a strong association between West African rainfall and intense hurricane activity along the U.S. Atlantic and Caribbean coastlines. His discovery is striking. Intense hurricanes, Saffir–Simpson categories 3, 4, and 5, occurred more frequently during the 20 years between the 1940s and the 1960s, compared to the 1970s and 1980s—except for 1988 and 1989. Figure 3 a depicts hurricane tracks during the period when West Africa was unusually wet, while Figure 3 b covers the 18 year period of severe drought, as in the Sahelian region of sub-Saharan Africa. Figure 4 similarly compares the ten wettest with the ten driest years in terms of hurricane intensity along the Atlantic seaboard and Caribbean areas. Apparently, as Gray remarks, "multidecadal climate cycles are a prevasive feature of the global ocean–atmosphere circulation, appearing in meteorological and oceanic data as far back as measurements have been made." He goes on to make clear that "the majority of Atlantic weather disturbances that eventually develop into intense hurricanes originate from West Africa." Thus it appears that the frequency of storms in the United States is inextricably

Table 4.2
Population Increases in Selected Coastal Cities

City	Increase (%)
Mississippi	2.0
Louisiana	0.8
Texas	19.6
Alabama	4.3
Florida	33.0
South Carolina	12.3
North Carolina	13.2

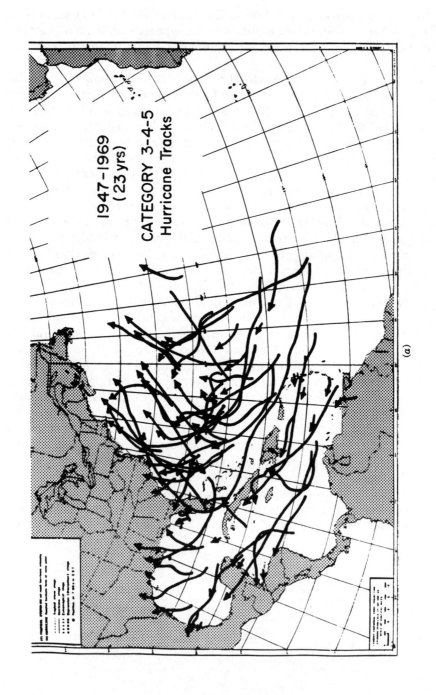

1947–1969
(23 yrs)

CATEGORY 3-4-5
Hurricane Tracks

(a)

178

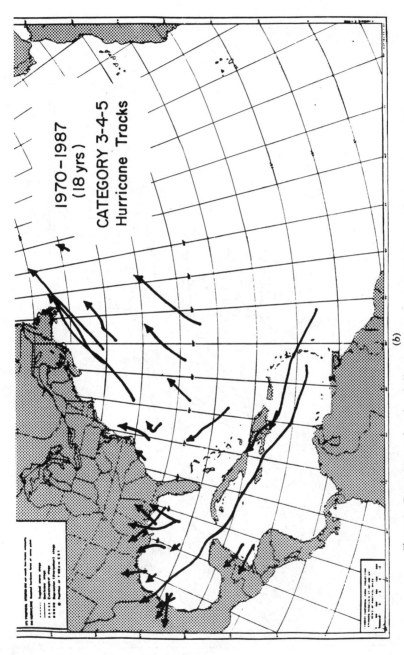

1970–1987
(18 yrs)

CATEGORY 3-4-5
Hurricane Tracks

(b)

Figure 3. Comparison of tracks of all major hurricanes (Saffir–Simpson category 3, 4, and 5) for (a) the 23-year period 1947–1969, when West Africa was wet, versus (b) the 18-year period 1970–1987, when it was dry.

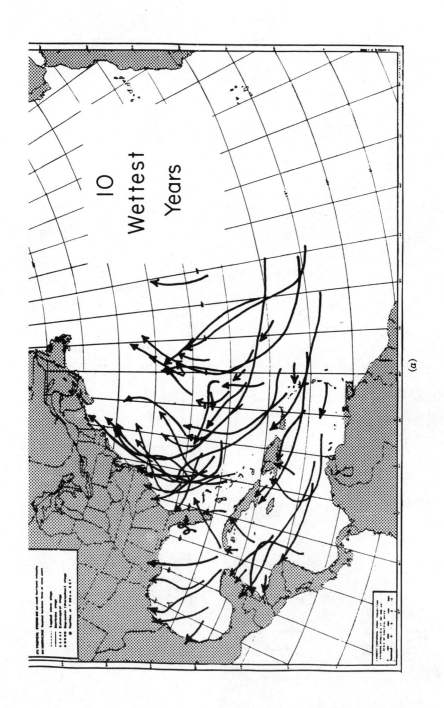

10
Wettest
Years

(a)

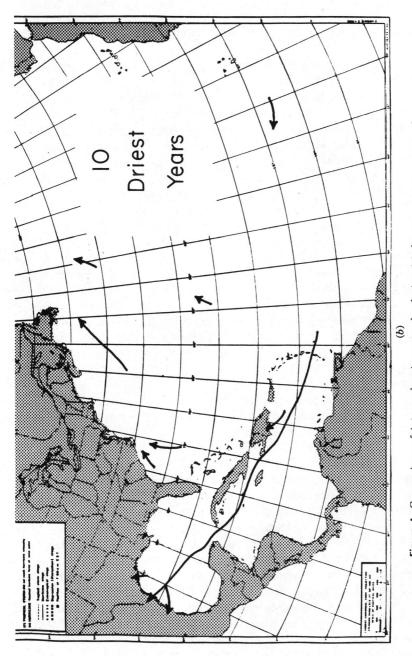

IO
Driser
Years

(b)

Figure 4. Comparison of intense hurricane tracks during (a) the ten wettest years in the western Sahel versus (b) the ten driest years during the 43-year period 1947–1989. (Courtesy of William M. Gray, Department of Atmospheric Science, Colorado State University, Boulder.)

tied to African weather patterns. If this is so, hurricane intensity and frequency over the coming decades may, in fact, be related to Sahelian weather rather than global warming. Is this the message in the exceptional hurricane activity for the years 1988 and 1989? Careful scrutiny is called for here.

Gray's data indicate that the relatively quiet decades are about over and that a more violent 20 years are in the offing or, if 1988 and 1989 are portents, have already begun. Separation of this decadal cycle from global warming, with its predictions of diverse climate dislocations, is of transcendent importance. Gray believes that, in "wet" years, changes in sea-surface temperatures induce development of a strong summer monsoon over West Africa, which in turn creates an environment favorable for low-pressure disturbances—"waves" that originate over West Africa and can evolve into hurricanes in the Atlantic. Figure 5 illustrates this concept. Prevailing winds are shown blowing eastward over West Africa. As they weaken, they move more slowly westward and become larger and organize into pre-hurricane clouds. In dry years, the waves are smaller and less organized and move faster and dissipate as they move across the ocean. Thus, as Figures 3 and 4 demonstrate, in the 18-year Sahel dry period—1970 to 1987—Hurricane Gloria was the only category 3–5 storm to arrive at the Atlantic seaboard. By contrast, between 1947 and 1969, 11 such storms hit the coast. Gray is also convinced that the 1990s and the early decades of the 21st century will be wet periods on the Sahel, and therefore intense hurricane activity should become a fact of life along our southeastern coast. Hurricanes Gilbert and Hugo may bear bitter testimony to his forecasts.

Perhaps his most evocative proposition, tucked away almost as an afterthought in his report, was the idea that "the historical data imply that such an increase in intense hurricane activity should be viewed as a natural change and not as a result of man's influence on his climate" (7). These multidecadal cycles are natural repetitive processes, which have been operating for at least hundreds of years and for which records are available. This requires appropriate attention, lest increased storminess be interpreted as evidence of CO_2-induced global warming. This will not be easy but, if true, the temptation must be resisted.

The elucidation of natural decadal cycles within cycles of longer time spans (i.e., 100, 1000, 20,000 years and longer) adds greatly to our knowledge base of both climate and weather, but it also warns us that interpretations of climate change forecasts must be done cautiously. For example, Gray's data suggest the following question: Does the fact of increased naturally occurring cyclone activity in the 1990s and into the 21st century preclude yet even more storms as a consequence of man-made atmospheric warming, with its increased evaporation from the oceans, such that the combination of the two would lead to even more severe storms? Furthermore, to make matters even more distrubing, Emmanuel reported that the

Setting the Stage for a Possible Period of Severe Storms

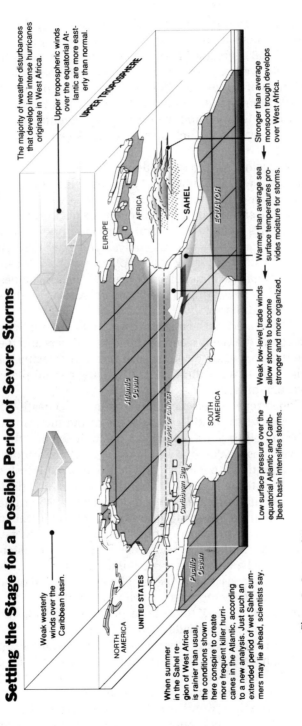

The majority of weather disturbances that develop into intense hurricanes originate in West Africa.

Upper tropospheric winds over the equatorial Atlantic are more easterly than normal.

Stronger than average monsoon trough develops over West Africa.

Warmer than average sea surface temperatures provides moisture for storms.

Weak low-level trade winds allow storms to become stronger and more organized.

Low surface pressure over the equatorial Atlantic and Caribbean basin intensifies storms.

Weak westerly winds over the Caribbean basin.

When summer in the Sahel region of West Africa is rainier than usual, the conditions shown here conspire to create more frequent killer hurricanes in the Atlantic, according to a new analysis. Just such an extended period of wet Sahel summers may lie ahead, scientists say.

Figure 5. Setting the stage for a possible period of severe storms. (Reproduced with permission of *The New York Times*.)

183

model he used to estimate the maximum intensity of tropical cyclones, given an atmosphere with the current CO_2 content, predicted hurricane forces with 40–50% greater kinetic energy than the most violent storms ever known. By way of comparison, Hurricane Gilbert, as noted earlier, contained winds of up to 179 miles per hour. Accordingly, this will seem relatively mild compared to winds of 225 mph. Such winds are far beyond Saffir–Simpson category 5 and would have the potential for unimaginable destruction, should one of such magnitude strike an exposed coastal seaport city.

Do tropical cyclones—the antecedent of hurricanes—operate as Carnot cycle engines, on which Emmanuel bases his calculations? To his credit, he offers a number of caveats and warns of the large uncertainties associated with climate simulations. In particular, he tells us that "the coupling of ocean dynamics to atmospheric circulation is presently quite crude." Nevertheless, he maintains that his analysis indicates that climate change "will lead to substantially enhanced topical cyclone intensity." If Gray's findings and Emmanuel's model are threads of the same tapestry, the next 20 years could be devastating to Caribbean countries and the southeastern cities of the United States.

Bangladesh, located on the northern bend of the Bay of Bengal, is mostly a low plain cut by the Ganges and Brahmaputra Rivers and their delta. It is high on the list of candidate countries at risk of increasing sea levels. The recent cyclone that swept through its low-lying coastal plain suggests its degree of vulnerability.

The Cyclone Preparedness Center has estimated a death toll of approximately 138,000, which could rise as some of the more remote areas are reached. On the other hand, as in past cyclonic events, the final figures could be lower as estimates give way to reality. But many more may die (are dying) as a consequence of cholera and lack of food and fresh water. Those who drink salt water will suffer further dehydration and death.

Not only was the initial response by the central government at Dahka slow and bungling, but once relief efforts were underway the infrastructure of roads, airports, and service stations to permit transportation of relief supplies by trucks or planes were discovered to be nonexistent. Nor could sufficient gasoline be found to assure the return of the vehicles once they had made it to the coast. Thus the problems were compounded, and the death toll unnecessarily increased. Cyclonic activity is not foreign to these shores. Since 1900 some 55 have been recorded. In fact, cyclones are as natural to Bangladesh as tornadoes are to Kansas.

Since the turn of the century, as many as 1 million people may have died as a result of high water and flooding. As population statistics generally are poor at best, the actual numbers will never be known. it hardly matters. The point is that the death toll is outrageously excessive.

What is needed is a rapid early warming system. But that requires trained meteorologists, computer experts, and a well equipped weather service—all expensive. But these will be ineffective if the response is nonexistent to poor—as it was on April 30, 1991, when winds of up to 145 miles per hour struck the area without warning in the early morning. And this event occurred at the early part of the cyclone season when the moon was full and tides were running normally high, which added an additional increment to the wall of water that enveloped the coast. In 1970, when over 300,000 people perished in the same general area, it came in the fall, toward the end of the season. Thus it is quite likely that before the 1991 season is over, the area will be hit again.

Additionally disturbing is the realization that it is almost impossible for any country, let alone a penurious one, to relocate 10 million people. Nor does Bangladesh have the billions of dollars needed to develop a network of barriers to the sea that would offer protection to its low-lying coastal plain. The fact that cyclones exact so great a toll here and lives continue to be lost in horrendous numbers has much to do with the fact that Bangladesh has existed only since 1971, and precariously at that, and prior to 1971, little was done when India and Pakistan were battling for control. Expectations of change for the better are not high.

CORAL REEFS AS CO_2 SINKS

Coral reefs are multicolored rocklike ridges formed from the hard outer skeletons of coral animals. Corals are marine invertebrates (no backbone) of the phylum Cnidaria, which include hydra, jellyfish, and sea anemones. All possess a saclike body with an opening to allow entrance of food and exit of waste. They are obviously primitive but have proved so successful in their environment that they are among the oldest continuously living things on earth, having thrived for at least 70 million years. Perhaps it is their sedentary life style that accounts for their longevity as they spend their lives fixed in place. Two types of coral animals are found on reefs: those that produce hard calcareous (calcium carbonate containing) shells or exoskeletons, and those that produce soft, proteinaceous skeletons. Only the calcium carbonate producers contribute to the formation of the limestone reef.

The nearest relatives of the coral are the anemones, with the common characteristic of a gastric cavity or a primitive stomach. Additionally, they exist only in polyp form, which means that their bodies are typically tubular or cylindrical, with one end having an oral cavity or mouth and surrounding tentacles and the opposite end fixed to the reef. Figure 6 shows a colony of the Scleratinian coral *Tubastrea* with their gastric cavities open and the tentacles aloft, while Figure 7 is a close-up view of the tentacles and gastric opening. All are carnivorous. Embedded in their tentacles are nematocysts, coiled and barbed stinging cells which, upon contact with a potential meal, most often microscopic plankton, instantly

Figure 6. View of a colony of the coral *Tubastrea* with closed oral cavity.

inject a venom that can paralyze and kill. Figure 8 is a close-up view of the fire coral *Millepora complanata*, showing released nematocysts at edges of the tentacles. As primitive as they are, they can be formidable creatures. Some species have been known to kill people.

The polyps build hard calcareous shells to protect their soft bodies. Young polyps absorb calcium and carbon dioxide from seawater and pro-

Figure 7. View of a single specimen of *Tubastrea* with gastric cavity open and tentacles spread.

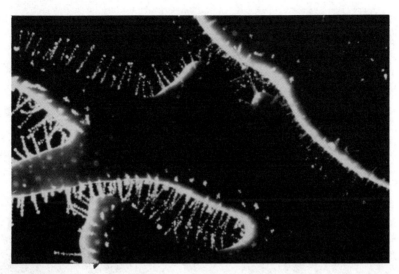

Figure 8. View of the fire coral *Millepora complanata* with trichocysts extended along its tentacles. (Courtesy of the Florida Marine Research Institute.)

duce the hard calcium carbonate skeletons, which also anchor the polyp in place. As they grow and die, side by side over thousands and hundreds of thousands of years, these multibillions of tiny creatures produce the ridges, which can be hundreds of feet thick and miles long. The largest in the world is the Great Barrier Reef, shown in Figure 9, located off the northeast coast of Australia, which extends over more than 12,000 miles of ocean, considered to be one of the largest, if not the largest, biological construction projects on earth. Its outer margins deflect and dissipate the energy of incoming waves, thereby protecting the coastal shoreline. Figure 10, taken from a glass-bottom boat, shows a section of the Great Barrier Reef, populated by a welter of coral species. At least a dozen types are visible.

The warm waters of the world support three types of reef: fringing reefs, which form along the edge of a tropical coast; a barrier reef, which is separated from a coast by a lagoon—the "barrier" is the protection afforded a coast from a pounding sea—and an atoll, which forms when a ring of coral develops on the sides of a submerged volcano. When the volcanic peak collapses and sinks into the sea, the ring of coral encircles a lagoon. As all reefs grow, they take the brunt of wave action, protecting coastal areas behind them. Over time, however, the constant pounding wears away the coral, producing a finely powdered material, which in large measure is responsible for the texture of soft, tropical, "sandy" beaches.

To say that a coral reef is a marine community is to describe a rose as a flower. Physically, a coral reef results from the growth of a broad spectrum of corals and symbiotic algae, animals, and plants. This growth creates an

Figure 9. Heron Island off the coast of Gladstone, Queensland, Australia, is part of the Great Barrier Reef. In the foreground is a channel cut through the reef to give tourist vessels access to the island. (Courtesy of the Australian News and Information Bureau.)

environment that, as Figure 10 depicts, becomes a habitat for literally hundreds of species of fascinating, multicolored marine forms—for example, fish, worms, crustaceans, and sponges. Reefs generally develop in relatively shallow water surrounding land masses in warm tropical seas. All are found in a relatively narrow band between the Tropic of Cancer and the Tropic of Capricorn, where water temperatures do not exceed 18–20°C. In addition to the warmer temperatures, reef growth requires high salinity, turbidity-free water. That's a key. Reefs would be "shot in the back," as it were, because rising seas, along with increased wind speed, would scour and erode coastal areas, which in turn would contribute solid particles of sand and gravel in such profusion as to create turbid conditions to a degree sufficient to block sunlight and thereby curtail photosynthetic activity. This would directly affect the zooxanthallae algae, which live in symbiotic relationship with the coral and contribute many of the organic chemicals needed for coral growth and development. Algae growth and reef development are further restricted to depths of less than 240–250 feet (80 meters), where light intensity is greatest. Consequently, rising seas would add greater depth, affecting the penetration of sunlight. The combined effects of added depth and increased turbidity are being viewed by some marine biologists as a death knell for coral reefs. In their terminology, rising sea level would produce "drowned" or dead reefs. Limestone (calcium carbonate) mounds would continue to exist but would be lifeless.

Reefs are sources of food and shelter for a wide diversity of marine animals and plants, but recently, human activity has placed great pressure on reefs. For example, the release of human and animal waste to the seas has spawned the overgrowth of algae, which block sunlight and bacteria, which consume oxygen. Together, these can be fatal to the coral animals and the reef. Along heavily populated coastal areas, the demand for additional farmland, roads, and living areas removes soil, which is washed into the sea, where it can form a sediment over the coral, burying and suffocating it. Thus over the past 30 years, reefs have come under heavy human pressure and global warming may continue the stress.

Curious as it may seem, destruction of the world's forests can also damage reefs. Removal of soil-binding vegetation leads to soil erosion. The

Figure 10 Seen through the glass bottom of a boat, this portion of coral reef is off the coast of Mackay, Queensland, Australia. At least six types of coral are readily discernible. (Courtesy of the Australian Information Service, New York.)

soil is carried by rivers and oceans as sediment, which can coat coral with a film that can kill rapidly. Such damage has already been observed in Hawaii, Australia, and Puerto Rico. Why is there concern for reefs? Any why are they so important? Recall that the vast structure of a reef is, in fact, a massive bulk of calcium carbonate, hundred of feet in height, built up and consolidated (cemented) against the forces of the sea by secretions of primitive plants and animals. Speaking of forces of the sea, it is noteworthy that reefs are at the mercy not only of human overactivity but of hostile natural forces as well. The El Niño event of 1987–1988 so warmed areas off the coast of Chile that the Galapagos reefs were all but totally destroyed. Estimates of the damage range up to 95%. And in the Caribbean, the magnificent reef off Punta Nazuc, Cancún, Mexico, was laid to waste by the onrush of Hurricane Gilbert. Is was not until after Gilbert left and the vast damage on land was assayed that scuba divers, returning to their favorite underwater haunts, discovered the carnage that Gilbert had visited upon the coral reef.

Australian researchers have estimated that all the world's coral reefs cover some 617,000 square kilometers of ocean while atolls cover another 115,000. Together, they produce about 900 million metric tons of calcium carbonate, which could raise to 1800 million metric tons as the reefs and atolls expand. Currently, they form a sink for about 111 millions tons of carbon—approximately 2% of the current emission of carbon dioxide. Thus they are prodigious sinks of CO_2. And water pollution notwithstanding, they have continued to grow, expand, and soak up CO_2. Should they drown and die, the CO_2 locked away in the limestone could be released into the atmosphere. The releases would be so staggeringly great as to make the contributions of CO_2 from burning fossil fuel pale by comparison. Perhaps that statement should be reread. Such releases would play havoc with climate—adding tremendous amounts of a greenhouse gas to an already taxed atmosphere. That is, of course, if temperature increases do indeed follow CO_2 increases.

Recently, a panel of marine biologists warned that environmental stresses, such as rising temperature and sewage runoff, were causing deterioration of coral reefs in the Caribbean as far north as the Florida Keys. Apparently, there is a growing problem of coral bleaching. Coral or the polyp without accompanying algae is white but becomes variously colored as a broad spectrum of algae colorize the coral. Microscopic plants, occurring by the many millions in the soft tissue, impart a blaze of brilliant colors—okra, brown, green, yellow, pink, lilac, burgundy, and mauve. Bleaching occurs when the algae die off or, for other reasons, do not grow, and without their growth, the coral dies, because the algae provide, as noted earlier, a chemical necessary for the laying down of the calcium carbonate skeleton. Although mass coral reef dying has been observed, officials at the National Science Foundation are not yet convinced that higher temperatures of the decade of the 1980s are related to the bleaching.

However, at the urging of concerned scientists, Congress may set up a coral reef alert center.

On the other hand, Australian marine biologists at the Great Barrier Reef Marine Park Authority estimate that, in the next 50 years, reefs generally should expand and, in the process, soak up a greater proportion of CO_2—as much as 4% in the 21st century. By way of example, the Great Barrier Reef, with an area of over 20,000 square kilometers, adds 50 million tons of calcium carbonate annually. They also see the eventual drowning of reefs, if the seas rise to levels predicted by current models. But reef drowning is not expected for many hundreds of years. Until then, reefs could continue to be major CO_2 sinks. Let us hope that the Australians are correct in their estimates and predictions.

TREES AND FORESTS AND THEIR REACTION TO CLIMATE STRESS

Although only one-third of the United States is forest land, many population centers are located close to forested regions, and forests are an integral part of American life. The fact that all forest lands are either owned by government or private interests suggests that all currently receive some type of management and that they would be a focus of considerable attention in a warmer world.

Unlike food crops and ornamental plants, forests—because of their size and life span—have not received the attention they deserve. Little information exists with which the evaluate the direct or indirect results of climate change on forest systems. Nevertheless, since temperatures and rainfall influence the distribution and composition of all biological systems, forests would be equally affected. Trees, utilize the C_3 photosynthetic pathway, which suggests that they would benefit from a CO_2-enriched atmosphere. On the other hand, paleobotanical studies indicate the shifting nature of forest boundaries under the stress of temperature changes. Recent studies suggest a northward expansion of most eastern tree species. The range of spruce, northern pine, and other hardwood trees could shift 60 or more miles over the next century.

Studies commissioned by the U.S. EPA to estimate how forests might respond to the warmer conditions predicted by global climate models consistently predict that the distribution of forests will reduced in their present ranges and that a slow migration northward is likely. Drier conditions would also mean less tree growth and therefore reduced forest productivity. But this estimate must be balanced by increased rates of photosynthesis and water use efficiency.

A critical question, for which there are currently few authoritative answers, is how quickly trees can migrate. The fate of many species could depend on the rapidity with which they shift to cooler regions, should current habitats become overly warm. Speed of migration appears to be the key to survival. If the climate models are anywhere near correct, by the year 2100, the earth will be warmer than at anytime over the past million years, but it is the rapidity of the forecast change that underlies the potential adverse effects. Over the next 100 years, average temperatures are predicted to rise by 2–5°C, a heretofore unimaginable increase. For many trees, a 1°C increase can mean a range shift of 60–70 miles—but only over a far longer time period. One hundred years may not be long enough for any type of tree to shift more than a few meters. As Margaret Davis of the University of Minnesota informs us, "the first lesson from the past is that species respond to climate individualistically. The fossil record documents the adjustments of range limits according to the unique sensitivities of each species to climate" (8). In this regard, Annabel Gear and Brian Huntley of the University of Durham, England, examined the record of change in the range of Scots Pine (*Pinus sylvestris*) and found that the 4000-year-old fossil stumps represented a period when climate changed abruptly and the pine shifted southward and became extinct, as a consequence of sudden cooling. They estimated that the range boundary shifted by 375–800 meters per year, which they believe is far too slow for survival of trees today, should global warming also occur abruptly—that is, in a hundred years or less (9). With so short a time frame for the difference between survival and extinction. Davis urges active human intervention in the form of planting and establishing temperature zone trees in locations appropriate for their growth into standing forests. Without such intervention, she predicts major losses of a broad variety of tree types. Thus she foresees the demise of sugar maples throughout their entire range, with the exception of the more northern areas of Maine and eastern Quebec. Probably the greatest losses would be sustained by beech trees, which currently range across large areas of the United States—as far south as Georgia and all the way north to Canada. Except for an area in Maine, Davis' model forecasts the wiping out of beeches throughout the country. Eastern hemlock and yellow birch await a similar fate.

Having heard the expression of an ecologist, it may be useful to consider the ideas of a horticulturist. S. H. Wittwer of Michigan State University tells us that "reassurances of the resiliency of agriculture and forestry to climate in the United States can be drawn from the present geographical range of crops. . . . Wheat, barley and potatoes," he points out, "can be grown throughout the United States, including Alaska. Soybeans are produced in almost every state east of the Rockies, and sunflowers can be grown from Texas to Minnesota. Important trees, such as Aspen, Red Maple, Douglas Fir and Ponderosa Pine, are found from Canada to Mexico. Making Minnesota the climate of Texas therefore would not eliminate many important crops" (10).

Given the great lack of knowledge about trees and forests generally, and their response to climate stress in particular, especially as it pertains to warming, model projections must be looked at with extreme caution. Thus it is to her credit that Davis, addressing a meeting of the Ecological Society of America, concluded by stating emphatically that "the message from paleoecology is clear, we cannot rely on analogy with the present to predict the future; we need functional understanding of the responses of individual species to multiple impacts. Larger scale models will have to build on the reactions of individual species to make accurate predictions of communities, ecosystems and landscape-scale biological systems. Global change taxes all our understanding of the dynamics of organisms and their interactions with the environment, in order to synthesize that knowledge into models that can predict ecological responses to combinations of environmental factors and rates of change that have never before impinged on the earth's biota." Here then is the essence of the issue. It strikes at the core of all attempts to divine the outcome to plants, trees, all vegetation, animals, insects, and minerals. The contemporary flora and fauna have never encountered the conditions predicted to occur. Davis' report was a clarion call for intensive research and reason. She had warned her audience at the outset to avoid temptation. "We must not build models that predict the future by shifting existing communities or biomes around. . . . If we do, we will make mistakes in predicting interactions with other components of the global system, because in the future new and different ecosystems are likely to come into existence" (8). Of course they will. Are the ecosystems that we know and have lived with for these past thousands of years the only ones acceptable? Could others be better? Are changes necessarily detrimental? Of course, new and different requires getting used to.

Earlier I noted that there was an ongoing debate between scientists as to the fate of the missing carbon—some 4.0 billion tons for which none of the scientific "bookkeeping" can account. Pieter Tans of NOAA believes there is an as yet undiscovered "sink" in the northern hemisphere, while others believe that the fertilizing effect of CO_2 and C_3 trees may be responsible for storing excess CO_2 in tree roots. Richard J. Norby of the Oak Ridge National Laboratory is testing these ideas by growing trees under enriched levels of CO_2. Thus far he reports beneficial effects on white oaks but only marginal increases in white poplars. Roots have yet to be examined for their content of CO_2. If this study proves successful, it would further advance the idea that large-scale planting of trees would be a relatively inexpensive yet appropriate means of soaking up excess CO_2.

AGRICULTURE AND CLIMATE STRESS

Climate variability has, from the day the first seeds were planted, made farming an uncertain business. That agriculture and climate are rhythmically linked is a given: farmers live by the weather. Thus predictions of a carbon dioxide-enriched world, with its forecasts of increased temperatures and alterations in patterns of precipitation, hold grave consequences for agriculture generally; food production could be a major casualty.

That the carbon dioxide content of the air is increasing is certain and well documented. What has not been widely remarked upon is the fact that carbon dioxide is one of the raw materials of photosynthesis, the process whereby plants use sunlight and CO_2 to make additional plant tissue—that is, more plants. Thus it is entirely possible that the more CO_2 is available to plants, the more productive they are likely to be. For many horticulturists and agronomists, unlike geophysicists and climatologists, adding more carbon dioxide to the atmosphere is tantamount to adding fertilizer. Commercial florists and growers of hothouse vegetables have known this for a century and have routinely doused their plants with carbon dioxide. Furthermore, it has long been heard that low levels of CO_2 in the air have been the factor limiting greater agricultural productivity worldwide. Thus contrary to broad opinion, the potential for enhanced productivity may be at hand. Interestingly enough, little has been said about the potential benefits that could accrue from an increase in the length of growing seasons. At this time, farmers may be the only ones unruffled by the idea of an atmosphere replete with carbon dioxide—having shifted from 270 ppm to 350 or more. In fact, among horticulturists, there is a growing belief that, for at least the past 40 years, there has been a detectable increase in the level of plant growth around the world, which may be related to that shift. Historical evidence suggests that the lower levels of atmospheric CO_2 are suboptimal for photosynthesis in the vast majority of food crops. The key term here is photosynthesis, which requires comprehension.

The fundamental task of photosynthesis is to make it possible for plant cells to convert inorganic CO_2 into carbohydrate, using energy absorbed from sunlight. In the early years of the 17th century, naturalists believed that plants derived their nutrients, and hence their growth, from the soil— a not unreasonable concept given the location of the roots. In a simple but elegant experiment, a Flemish physician, Jan Baptista van Helmont (1577–1644), disproved this idea. He planted a 5 pound willow tree in a box containing 200 pounds of oven-dried soil. After 5 years on a diet limited to rainwater as needed, the tree had gained 164 pounds and the soil was 2 ounces shy of its original 200 pounds. Given that water was the only substance added, van Helmont could hardly be faulted for concluding that it was the water, not the soil, from which the tree drew sustenance. He was mostly wrong, but not totally. Minerals in the soil, conveyed by water to the roots, do play a role. But it would take another 300 years before contemporary scientists would demonstrate that the primary plant nutrient was inorganic carbon dioxide, which plants obtain from the air surrounding them. Thus it is carbon dioxide that is the essence of the process of photosynthesis, and it is upon photosynthesis that life on earth is based. For a chemical present in the global atmosphere at such remarkably low levels, carbon dioxide is probably the earth's most important substance. Without it, all living things would die. Plants cannot produce more of themselves, and without plants, animals and insects and other related species along the food chain die with them. Without plants and animals, people must perish. The importance of carbon dioxide and its precursor carbon simply cannot be overstated.

In its most elemental form, the process of photosynthesis can be stated as

$$CO_2 + H_2O \rightarrow C_6H_{12}O_6 + O_2,$$

which means that the inorganic gas carbon dioxide, together with water (somehow), produces an organic compound: in this case, the six-carbon sugar glucose and oxygen. Its simplicity raises more questions than it answers. An improved statement would be

$$sunlight + 6CO_2 + 6H_2O \rightarrow C_6H_{12}O_6 + 6O_2 + 686 \text{ calories},$$

which indicates the need for light energy to drive the reaction, the number of molecules of CO_2 and water needed to produce a molecule of carbohydrate, and the fact that the reaction produces heat. But this is far from satisfactory for plant biochemists or those chemical engineers who, as we shall later see, look on photosynthesis as a potential source of clean fuel. These people would prefer a statement on this order:

$$6 \text{ ribulose-1,5-diphosphate} + 6CO_2 + 18ATP + 12H_2O + 12NADPH + 12H^+ \rightarrow$$

$$6 \text{ ribulose-1,5-diphosphate} + glucose + 18P_i + 18ADP + 12NADP.$$

The enzyme ribulose-1,5-diphosphate is placed on both sides of the equation to show that it is a necessary component for the reaction to go forward and is regenerated at the end of the reaction. As we shall see, ATP (adenosine triphosphate), ADP (adenosine diphosphate), NADPH (nicotinamide-adenine dinucleotide phosphate hydride), and NADP (nicotinamide adenine dinucleotide phosphate) are energy-bearing chemicals essential for the actual conversion of CO_2 and H_2O into carbohydrate. Figure 11 illustrates the basic physicochemical process. Not shown in the figure are some 15 essential chemical transformations that ultimately yield the carbohydrate and oxygen. But the figure does show the thylakoid membranes inside the chloroplasts where these reactions occur. Along with the obvious importance of CO_2, it is essential to recall that oxygen, without which there would also be no life, is another major product of photosynthesis. In a remarkable photographic achievement, the release of oxygen from a marine plant is shown in Figure 11 in Chapter 2.

All plants use one of three metabolic pathways to produce carbohydrate. Thus their responses to increased levels of CO_2 depend, in large measure, on which pathway is used. Photosynthetically, plants are classi-

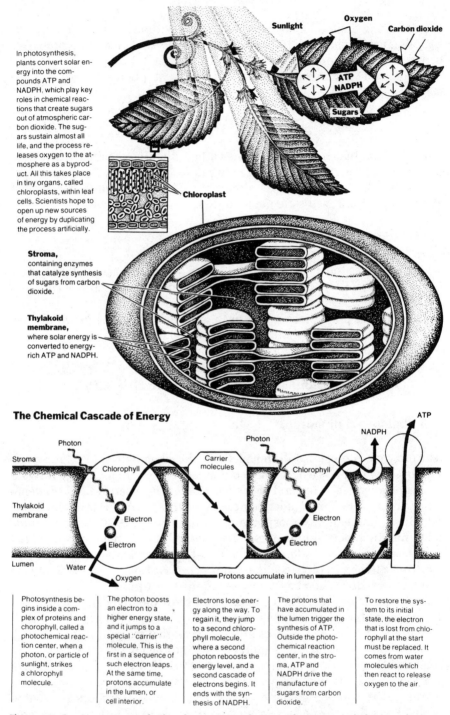

In photosynthesis, plants convert solar energy into the compounds ATP and NADPH, which play key roles in chemical reactions that create sugars out of atmospheric carbon dioxide. The sugars sustain almost all life, and the process releases oxygen to the atmosphere as a byproduct. All this takes place in tiny organs, called chloroplasts, within leaf cells. Scientists hope to open up new sources of energy by duplicating the process artificially.

Sunlight   Oxygen   Carbon dioxide   ATP NADPH   Sugars

Chloroplast

Stroma, containing enzymes that catalyze synthesis of sugars from carbon dioxide.

Thylakoid membrane, where solar energy is converted to energy-rich ATP and NADPH.

The Chemical Cascade of Energy

Stroma   Photon   Chlorophyll   Carrier molecules   Photon   Chlorophyll   NADPH   ATP

Thylakoid membrane   Electron   Electron   Electron   Electron

Lumen   Water   Oxygen   Protons accumulate in lumen

Photosynthesis begins inside a complex of proteins and chorophyll, called a photochemical reaction center, when a photon, or particle of sunlight, strikes a chlorophyll molecule.

The photon boosts an electron to a higher energy state, and it jumps to a special "carrier" molecule. This is the first in a sequence of such electron leaps. At the same time, protons accumulate in the lumen, or cell interior.

Electrons lose energy along the way. To regain it, they jump to a second chlorophyll molecule, where a second photon reboosts the energy level, and a second cascade of electrons begins. It ends with the synthesis of NADPH.

The protons that have accumulated in the lumen trigger the synthesis of ATP. Outside the photochemical reaction center, in the stroma, ATP and NADPH drive the manufacture of sugars from carbon dioxide.

To restore the system to its initial state, the electron that is lost from chlorophyll at the start must be replaced. It comes from water molecules which then react to release oxygen to the air.

Figure 11. Green power: inside the plant's energy factory. The process of photosynthesis is detailed physicochemically. (Reproduced with permission of *The New York Times*.)

fied as C^3, C^4, and CAM. The enzyme ribulose diphosphate carboxylase catalyzes the insertion of CO_2 into the metabolic pathway early in the process to form the three-carbon compound 3-phosphoglycerate, hence the designation C^3 for all plants that use this pathway to carbohydrate formation. Another smaller group inserts CO_2 into the metabolic pathway via the enzyme phosphoenolpyruvate (PEP), which yields the four-carbon intermediate oxalacetate; thus their designation C^4. CAM, or crassulacean acid metabolism, is limited to desert plants, that is, succulents that have evolved in arid conditions. In so doing, they have adapted to low moisture levels by being able to keep their stomata closed in the hot, dry daytime when water loss would normally be great. Their stomata open at night, taking in CO_2 when it is cooler. Of the major plant crop, only the pineapple follows the CAM pathway. About 5% of all plants derive their carbohydrate for the C^4 pathway, while another 95% are C^3's. Table 4.3 lists the major crops by their respective metabolic pathway. The importance of the C^3 and C^4 response to CO_2 enrichment underlines the current concern for world food supplies.

In field and laboratory trials—using growth chambers and greenhouses—C^3 plants exhibit a far greater response to added CO_2. It is almost as though they have been starved of it. In fact, the C^4 plants evolved fairly recently in geologic time and in an atmosphere of naturally reduced CO_2. The C^3 plants, on the other hand, are older species that evolved at a time of naturally higher CO_2 levels. Numbers of scientists have suggested that these crops have been functioning in a suboptimal CO_2 environment, which limited their productivity. Thus their favorable response to doubled or trebled CO_2 should come as no great surprise. They seem to respond to CO_2 as they might do to fertilizer. Evidently, high or higher CO_2 levels are optimum for their development. C^4 plants, on the other hand, produce carbohydrates more efficiently and thus are more productive under the existing atmospheric regime. Based on early experimental field trials, as much as a 10% increase in growth and yield of C^4 crops such as maize and sugar cane can be expected, for a doubling of CO_2 from 350 to 700 ppm. For C^3 crops such as wheat, soybean, and rice, increases in yield of 10–50% have been obtained—depending on growing conditions.

Table 4.3
Examples of C^3 and C^4 Crops

C^3 Crops	C^4 Crops
Wheat, barley, rice, soybean, alfalfa, cassava, cotton, potato, sweet potato, rye, oats, beans, sugar beets, tomato, banana, orange, grapes, coconut, mango, onion, watermelon, cucumber, eggplant, carrots, and all trees	Corn, sorghum, millet, sugar cane, and maize

Bruce Kimball of the U.S. Water Conservation Laboratory, Phoenix, Arizona, reviewed the published literature on the effects of CO_2 concentration on agricultural yield. From over 700 studies on the yield of 38 crops, he concluded that increasing the CO_2 concentration from 330 to 660 ppm would likely lead to an increased yield of about $32 \pm 5\%$. This level of increase has since been affirmed and substantiated (11).

In one such experiment, still in progress, sour orange trees—*Citrus aurantium*—have been exposed, albeit in a controlled environment, to almost double the ambient level of CO_2, with fruitful results. Figure 12 contrasts the difference in growth response between two orange trees. The one on the left is growing in ambient air, while the one on the right is growing in air supplied with an additional 300 ppm of CO_2. At the time that this photo was taken (August 1990), the trees had been growing in these differential CO_2 environments for over $2\frac{1}{2}$ years. The background is worth reviewing, and this case can act as a model or example of what is happening around the country and world with many other essential crops.

Figure 12. Sherwood B. Idso of the USDA's Agricultural Research Service stands before two sour orange trees growing within clear-plastic-wall, open-top chambers at Phoenix, Arizona. The smaller tree is growing in ambient air and the large one in air supplied with an extra 300 ppm of CO_2. At the time of this photo, August 1990, the trees had been growing in these differential CO_2 environments for over $2\frac{1}{2}$ years. (*Courtesy of the USDA.*)

In July 1987, eight 30-cm tall sour orange tree (*Citrus aurantium*) seed-lings were planted directly into the ground at Phoenix, Arizona. Four identically vented, open-top, clear-plastic-wall chambers were then con-structed around the young trees, which were grouped in pairs. CO_2 en-richment—to 300 ppm above ambient—was begun in November 1987 in two of these chambers, continuing unabated to the time of this writing. Except for the differential CO_2 enrichment of the chamber air, all the trees have been treated identically, having been irrigated at periods deemed appropriate for normal growth and fertilized by procedures standard for young citrus trees.

During the months of January and February 1990, the diameter of the trunk and every branch of every tree were measured at 20-cm intervals along their entire lengths. Converting these data into volumes, Idso found the total trunk and branch volume of the CO_2-enriched trees to be 2.79 times greater than that of the ambient-treatment trees.

In July 1990, a comprehensive assessment of the trees' fine-root biomass was made. Summed over the entire 120-cm soil depth investigated, the CO_2-induced enhancement of root growth resulted in a 75% increase in fine-root biomass beneath the canopies of the trees. Because the roots of the CO_2-enriched trees extended farther out from their trunks than did the roots of the ambient-treatment trees, the CO_2-enriched trees had 2.75 times more total fine-root biomass than the trees grown in ambient air.

The researchers at the Water Conservation Laboratory raised a very pertinent question: How can sour orange trees—or any trees, for that matter—achieve a near-tripling of both above- and below-ground biomass with a less-than-doubling of the atmospheric CO_2 content? In attempting to answer this question, they measured the net photosynthetic rates of the trees between 7 a.m. and 5 p.m. on six clear days throughout the 1989 growing season and found that the mean net photosynthetic rate of the CO_2-enriched trees was 2.36 times greater than that of the ambient-treatment trees over that period. Then, on several clear days throughout the summer of 1990, they made similar measurements over a number of 24-hour periods, finding that the nighttime respiration rates of the CO_2-enriched trees were approximately 30% less than those of the trees growing in ambient air. Combining these two effects showed that 24-hour CO_2 sequestering in the CO_2-enriched trees was approximately three times greater than it was in the ambient-treatment trees.

Idso's interpretation of the results are worth considering. "The signifi-cance of our results and the implications they portend for other trees is perhaps best illustrated by a set of calculations performed by Dr. Gregg Mariand of Oak Ridge National Laboratories (1988). He concluded that a doubling of the growth rates of the world's existing forests would be more than sufficient to absorb all of the CO_2 that is yearly released to the atmosphere by the activities of man, and that maintaining that growth advantage for only a little more than three decades would return to the biosphere all of the carbon that has been released to the atmosphere over the past two centuries. Our experiments demonstrate that this scenario is well within the range of what may be considered likely, and that the CO_2 'problem' may thus hold within it the seeds of its own solution (12)."

Obviously, the response of what otherwise would be commercial field crops to controlled enrichment is unlikely to be as favorable as the response of "pampered greenhouse crops," for a welter of reasons, including lack of water or nutrients, unfavorable temperatures, and competition. Attempting to predict what will occur in the real world, using laboratory-gathered data, contains the stuff of controversy. For example, let us suppose that with CO_2 enrichment plants produce greater yields per acre. Will the fruit or vegetables be as tasty? Will starch and/or sugar content be diluted? The investigators who are concerned with yield probably have not considered organoleptic qualities. Some researchers suggest that fallen leaves may not decompose as readily, thereby slowing release of nutrients back into the soil. But others maintain that increased earthworm ability would offset the loss. There is sufficient difference of opinion to keep laboratories running around the clock. The differences of opinion, however, appear to be of a reasonable nature that requires additional study. It must be recalled that if CO_2 enrichment enhances growth of both C^3 and C^4 plants, it will also enhance weed growth. If plants that have an adaptive edge benefit from increased CO_2, then weeds may well overcome their nonweed neighbors. If this should occur, weed control could take on greater importance. Wittwer has noted, however, that of the 20 most important food crops, 16 are C^3 types, while of the world's most annoying weeds, 14 are C^4. As noted earlier, C^3 plants respond more robustly to added CO_2 compared to C^4 plants. In tests of the competition between species of both C^3 and C^4 plants have held their own. Thus for the 16 food crops, weeds may just be less of a problem than they currently are. Conversely, corn, millet, sorghum, and sugar cane, which are C^4 plants, are likely to feel the competition from C^3 weeds (13).

Experiments additionally confirm the observation that high CO_2 levels alleviate water stress in plants. In growing areas where model projections suggest declines in precipitation, greater intrinsic water efficiency would offset potential losses. Thus, for example, yields of water-stressed wheat grown at high CO_2 levels appear to be as large or larger than well watered wheat growing at ambient CO_2 levels. Evidence is also at hand that high CO_2 levels increase as plants use water efficiently. Water needs to both absorb and lose heat before its temperature changes to any appreciable degree. That is, water has a high specific heat capacity, which protects tissue cells from excessive changes in temperature—up to a point. Water is also transparent, allowing sunlight to penetrate and reach the chloroplasts where light energy, as noted in Figure 11, drives photosynthesis. Most of the water that a plant loses via evapotranspiration escapes from the stomata, the cells that line the epidermis. During the day, stomata, as shown in Figure 13, open in response to solar radiation. Carbon dioxide enters through the stomata and diffuses into the leaf through the cells lining the pore. It is from here that the CO_2 moves to the chloroplasts, where it is converted to carbohydrates in the photosynthetic process. At the time that CO_2 enters the stomata, moisture diffuses out to the drier air around the leaf. This loss of water, for the most part unavoidable, initiates the flow of water from the soil through the plant, to the evaporative sites. To some degree the plant can control the rate of transpiration by varying the extent to which the stomata are opened. In sunlight, the plant maintains tight control on the degree to which the stomata open, responding to the dryness of the air and the water potential of the leaf. Of particular importance is the movement of water from the soil to the plant, a vital and complex process. But to what degree the plant can control this movement will depend in large measure on the environment. Changes in the surrounding environ-

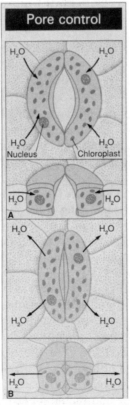

Figure 13. Stomata open when water moves into the guard cells and makes them turgid. (A) The thick inner sides of the guard cells bow outward, opeing the pore. (B) When the concentration of mineral ions is low, water moves into the surrounding cells and the guard cells begin to collapse. (Reproduced with permission of IPC Magazines, London, England.)

ment could disrupt the process at any point. So, for example, a higher air temperature can increase the potential for transpiration. How the combination of warmer air and higher CO_2 will affect transpiration in the fields, as it were, remains to be seen. Nevertheless, laboratory studies do suggest that high CO_2 in conjuction with increased temperature has positive, beneficial results. Here again Wittwer may be on to something. He informs us that during the past 30 years there has been a great increase in crop productivity and he ascribes this to the "more efficient and extensive use of water, greater efficiency of photosynthesis, improved biological nitrogen fixation, remarkable genetic improvements, and the extensive use of mechanization, pesticides, growth regulators, and fertilizers." These gains in productivity have also coincided with the period of most rapid increase in concentration of atmospheric CO_2. "It is tempting," he suggests, "to ascribe some of this increase in productivity of the past and present to the rising level of atmospheric CO_2" (13). Perhaps it would bode well for the future food supply, depending on the concurrent effects of increased temperature and moisture. But again, until widespread field trials are undertaken, uncertainty will continue to dog all discussion. But a warmer, wetter world can have a downside.

PLANT DISEASES AND CLIMATE STRESS

Cereal grains are highly susceptible to fungal infections. Where the responses of plants and trees to climate change are speculative, if not problematical at this time, the potential for ill effects of increased moisture on cereal grains is well known. Ergotism, St. Anthony's Fire, has a long and fearsome history. It was the great terror of the Middle Ages (the 9th–15th centuries) and was referred to as *mal des ardents*, sickness of the burning and scorched, referring to the feeling of being on fire, and it was called St. Anthony's Fire because only supplication to him was supposed to have quenched the fire. At the time, entire villages became psychotic almost at a signal. But ergotism was well documented in the American colonies and central Europe and was especially well known in Russia.

Ergotism is a bizarre affliction characterized by uncontrolled, deranged behavior with convulsions and hallucinations, the consequence of eating bread and related products prepared from flour derived from grain contaminated with the fungus *Claviceps purpurea*. The spurlike ergots, the sclerotia shown in Figure 14, consisting of the dark, hard, compact mass of hyphae, contain a mixture of highly potent, pharmacologically active alkaloids—nitrogenous ring compounds such as ergotamine, a powerful vasoconstrictor and muscle contractant once used to induce labor in pregnant women, along with the powerful hallucinogens lysergic acid and lysergic acid amides. Perhaps the use of the letters LSD will have greater meaning. But we need not harken back to the Middle Ages for instruction.

Figure 14. *Claviceps purpurea* is a notorious fungal parasite of grain. The spurlike, hard, dark, compact masses of hyphae growing on the grain are the toxin-containing ergots. (Courtesy of Robert W. Goth, Agricultural Research Service, USDA.)

Far closer in time was the harrowing experience of a number of villages in France in August 1955. At that time, masses of villagers of Point Saint Esprit, Connaux, Bagnols sur Ceze, and Genies de Conolas, Rhone Valley hamlets of Provence and Languedoc, and the Midi region of France were stricken by a violent form of food poisoning—an epidemic of ergot/LSD-induced psychosis. Flour contaminated with huge doses of ergotamine were baked into loaves of baguettes, flutes, and gros pain. Within hours of eating the contaminated breads, people experienced stomach cramps and diarrhea, and within 18–24 hours deranged behavior was noted. Descriptions of two crazed individuals may serve to portray the insufferable effects of the contaminating alkaloid. But there was mass hysteria among the people not stricken, as they watched their loved ones and neighbors suddenly become violent and implacable strangers. The overwhelming fear of not knowing where and when it would strike next dominated everyone's thoughts. A normal person one minute could be a demented maniac the next. That description fit Charles Veladaire, who suddenly broke away from a group of friends and ran down the Rue Beauregard toward the Rhone, his startled friends chasing him, calling to him to come back. At the Quai de Luynes he stopped on the edge of the river and warned his friends to stand back. "Don't touch me! Stand back all of you," he yelled. "I am dead, do you hear? I am dead! My head is made of copper, and I have snakes in my stomach! They are burning, burning, burning!"

He turned to face the whirling current of the Rhone, which was dark, ominous, and turbulent beneath him. Two of his friends approached him cautiously.

"Stand back! Don't come near me!" he yelled again. "The snakes are burning me, don't you understand? They'll burn you to death!"

He turned again and prepared to jump. One friend, moving swiftly, tackled him at the knees. The other, rushing to the side, pulled both of them back as they grappled on the edge of the stone embankment. All three were poised precariously until the rest of the group could reach them. The scuffle continued until it moved back on the darkened quay under the pressure of the mass of men. There were seven others, and even then they could barely hold Valadaire down. He broke away once and dashed madly toward the bank again, but he was trapped and held down until his struggling subsided. On his feet, they guided him along the Rue Saint Jacques and the Boulevard Carnot to the hospital. Twice he broke away again, and twice he was recaptured, bruised and bleeding.

Joseph Puche had a more fantastic response. "I can fly! Don't all of you believe me?" he yelled. "Watch me!"

All of this had taken less than 10 seconds. With no further warning, he jumped, arms outstretched as if they were the wings of an airplane.

He soared through the air, landed on the Boulevard Carnot, almost telescoping both his legs. He screamed in agony. Dr. Gabbai ran to his side. The sisters ran out of the hospital. As the doctor reached his side, the blood was already soaking through his hospital gown from the greenstick fractures. Then, incredibly, he rose up on his shattered legs and ran full speed down the boulevard. Several stunned villagers, from the houses around, joined the chase. He was not caught until he had run 50 meters on two broken legs. It took eight men and the doctor to drag him, still protesting that he was an airplane, back to the hospital. It took all eight to hold him down as the doctor set the fractures. The plaster casts were applied from the hip down to the toes. Immobilized by the casts, he thrashed his arms wildly about the bed until exhaustion finally conquered him.

Then the grotesque and macabre parade began. Mixed with those who were genuinely stricken with wild and uncontrollable psychoses were those who were victims of emotional hysteria, so great was the fear throughout the village. The hysterical would have to wait; the psychotics would require all the resources of the hospital and more (14).

High moisture and warmer temperatures provide optimum growth conditions for a number of fungi, but especially *Claviceps*. The Middle Ages with its years of excessively wet weather and often warm springs and summers were prosperous times for this mold. Chroniclers at the time wrote of the epidemics of madness in England with people seeing demons everywhere, and in Germany of outbreaks of hallucinations and convulsions in the Rhineland. In 1482, for example, the chronicles tell of people bashing their heads against stone walls and running into the river and dying in agony. Among modern-day historians and correlation between excessive wet weather and growth of *Claviceps* is striking. According to Mary Kilbourne Matossian, "the best predictor of epidemics in the British Isles during the period 1350–1489 was summer wetness"(15).

Available moisture is a limiting growth factor. Fungi do not grow on grain with moisture content below 12–13%. Consequently, drying cereal grain prior to storage can be a primary preventive measure. But grain that is infected while growing does not benefit from later drying. The mycotoxin is heat stable.

Cloudy, wet springs and summers favor infestation by lengthening the time that rye flowers remain open and vulnerable to infection with *Claviceps* spores. As the mold grows, its content of alkaloid increases; the level of production appears temperature dependent. Higher spring and summer temperatures induce more potent alkaloids. The insidious fact is that when the flour made from contaminated grain is kneaded into dough, there is no off-odor or off-taste. As was well documented in the outbreak in France, the rising of dough was unaffected and the finished loaves could not be distinguished from uncontaminated loaves. However, when ergotamine-containing dough is baked, the 350°F oven temperature converts ergotamine ($33H_{35}N_5O_5$ to $C_{16}H_{16}N_2O_2$—LSD-25). In Port Saint Esprit, the doses were massive. Three died and 50 became insane; others recovered after severe physical injuries and distressing bouts of violent nightmares. More recently, a range of medical studies have demonstrated that in small doses LSD and lysergic acid can induce such harmful long-term biochemical injury as chromosome damage and inhibition of immunoglobulin formation, as well as increasing the risk of spontaneous abortion. The potential for harm is real and serious.

Claviceps and ergotism could become widespread again if excessive rainfall and warm temperatures become the dominant conditions in cereal growing areas. Careful inspection of grain would be required to prevent contamination of flour. It would be difficult to imagine a totally effective inspection system.

THE GREAT FAMINE: A RESULT OF CLIMATE STRESS

Phytophtora infestans (Figure 15) is another moisture-loving mold. This one has a predilection for potatoes and was responsible for devastating potato crops in Europe and Ireland during the 1840s and 1850s. The

(a)

(b)

Figure 15. Two views of *Phytophtora infestans*, the blight of potatoes. (Courtesy of Robert W. Goth, Agricultural Research Service, USDA.)

combination of wet weather and this fungus was responsible in large measure for 1 million deaths in Ireland and the subsequent waves of migration that followed in its wake.

Intensive potato cultivation in Ireland had accommodated a remarkable 70% population increase between 1745 and 1841. By 1840, potatoes were the sole food of 30% of the people and essential to another 30%. Potatoes were eaten in place of bread and the two main meals of a working person's day consisted of potatoes.

Blight first occurred in autumn of 1845. In 1846 and 1848 the crop failed entirely. On August, 1845, John Lindley, the newly appointed and first professor of botany at the University of London and editor of the *Gardeners Chronicle and Horticultural Gazette*, wrote: "A fearful malady has broken out among the potato crop. On all sides we hear of destruction. In Belgium the fields are said to be completely desolate. There is hardly a sound sample in Convent Garden market. . . . As for cure for this distemper, there is none." And he continued: "We are visited by a great calamity which we must bear." The desperate nature of the devastation was announced on September 13th. "We stop the press with very great regret to announce that the potato Murrain has unequivocally declared itself in Ireland. The crops about Dublin are suddenly perishing. . . . Where will Ireland be in the event of universal potato rot?" (16). The rot was all encompassing. Starvation was widespread. A wholesale exodus followed in which hundreds of thousands fled the country. Between 1845 and 1851, the population of Ireland declined by more than 2%. A mold—a fungus— changed the course of Irish history.

YELLOW RUST

Recent increases in both frequency and severity of outbreaks of *Puccinia striiformis*- induced yellow rust of winter wheat have been reported from the northwestern United States. These increases in losses of winter wheat crops have been laid to changes in climate. Milder winters have helped the survival of the fungus and shortened its latency period. But colder springs reduced the wheat's disease resistance, rendering it more susceptible to mold infection. Yellow rust presents itself as elongated yellow-orange streaks on affected plants with nodular clumps running their length, which burst regularly, releasing spores that continue the infection. The result is lowered crop yield. Crop yields can also be appreciably reduced by the Colorado potato beetle (*Leptinotarsa decemlineata*), which feeds unmercifully on potato leaves. Its number wax and wane with changes in climatic conditions. These few examples, which could literally be multiplied a hundredfold, personify the great potential for ravaging our food supply by microbial and insect pests whose life cycles are finely tuned to temperature and moisture.

There is, however, yet another consideration that could well work to mitigate the potentially disruptive effects of excessive microbial and arthropod growth. From laboratory studies it is known that increasing carbon dioxide levels cause constriction of plant stomatal openings, and from both laboratory and field studies it is well established that microbial and insect pathogens gain entry to plants through their stomatas. For those plant pathogens that cannot penetrate intact plant tissue, the stomata offer portals of entry. It is also known that as CO_2 levels are increased, stomatal openings constrict. Accordingly, in a high or higher CO_2 world, it may be increasingly difficult for pathogens to infect crop plants. The increased temperature and moisture favorable to pests may be neutralized by increasing CO_2 levels. The incidence of plant disease could well decrease with higher CO_2 levels, and with a decrease there would be a parallel increase in per acre yield. Obviously, the issue is neither all black nor all white.

HEAT SHOCK PROTEINS: RESPONSE TO TEMPERATURE INCREASES

High temperature is an environmental stress for plants and causes the activation of specific genes, resulting in the rapid synthesis of a group of polypeptides referred to as heat shock proteins(HSPs). There HSPs are a protective mechanism that allows plants to survive exposure to what would ordinarily be lethal temperatures. But survival, especially in cereal grains, does not necessarily equate with optimum food quantity—that is, the use that humankind makes of these crops.

At the Australian CSIRO's (Commonwealth Scientific Industrial and Research Office) Division of Plant Industry's Wheat Research Unit, Caron Blumenthal and her colleagues have recently found that weaker dough strength results from grain subjected to temperatures over 30°C (86°F) during ripening in both greenhouse and field trials. They have found HSPs in roots, shoots, leaves, embryos, and endosperm of wheat plant tissue. And they have determined that heat shock spurs the synthesis of gliadin, a gluelike protein that reduces dough strength. It appears that temperature fluctuations over 35°C for only a day or two were sufficient to reduce wheat quality. Figures 16a describes dough strength (Rmax) in relation to hours above 35°C, and Figure 16b displays the years between 1960 and 1987 with temperatures above and below 35°C. It is evident that weaker dough strength for hard wheat coincides with episodes of elevated temperature. Note too that, over the past 30 years, those growing seasons when dough strength (Rmax) was lowest were also the years with the warmest ripening periods—that is, the number of hours above 35°C. Blumenthal and co-workers have developed a hypothesis—set out diagrammatically in Figure 17—which provides a comprehensive overview of the process (17, 18). Here the link between response to heat shock in the field and final weakening of dough strength is clearly set forth. The elegance of this research is not only that it has uncovered an underlying plant defense mechanism, but rather that it points the way to assessing trials of a variety of species of wheat which will ultimately lead to the development of a genotype with tolerance to heat stress—one whose dough strength will remain strong and

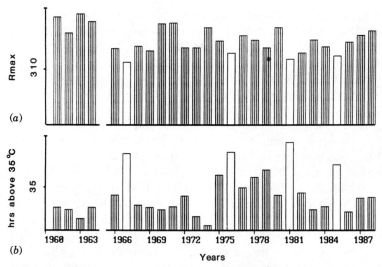

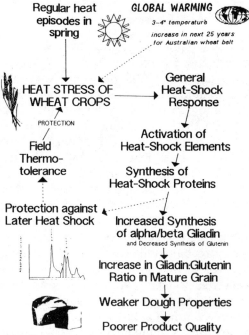

Figure 16. Weaker dough strength for Australian prime hard wheat coincides with high-temperature episodes. During the past decades, those seasons when dough strength (Rmax) was least were the ones with the hottest ripening periods—hours above 35°C—from October 1 to December 12. (Courtesy of Caron Blumenthal, CSIRO, Divsion of Plant Industry, New South Wales, Australia.)

Figure 17. Links in the heat-shock response to thermotolerance in the field and to the weakening of dough strength after heat stress. (Courtesy of Caron Blumenthal, CSIRO, Division of Plant Industry, New South Wales, Australia.)

organoleptically acceptable under conditions of prolonged field heat. Should global warming increase the frequency of days with excessively high temperatures, as is currently being forecast, bread will not become a casualty.

Research on HSPs is not limited to dough quality. Bacteria and a great diversity of higher plants are known to respond to temperatures beyond their normal range by producing HSP. In fact, a response can occur within 10–20 minutes of exposure. Once produced, the proteins confer an ability to prosper in hot environments. Researchers at the International Crop Research Institute for the Semi-Arid Tropics in Hyderabad, Pakistan, in conjunction with scientists at the Welsh Plant Breeding Station have developed a process that permits testing for HSP content in sorghum, the major cereal grain of over half-a-billion people in the tropics. When soil temperatures exceed 50°C (122°F) young sorgum plants wilt and die, producing serious food shortages for the local populations. Now with a relatively simple test for the presence of HSPs, botanists will be able to quickly select heat-resistant plants for widespread planting—which will help stabilize world food supplies. With temperatures in the tropics expected to rise even by only a small amount, discovery of HSPs and rapid analysis for them must be considered a major, though as yet little known, contribution.

AFLATOXICOSIS

There is an obverse side to plant infestation by microbial pathogens; high temperatures and drought. *Aspergillus flavis* is a fungus that thrives on corn, wheat, peanuts, and rice, which have been weakened by prolonged periods of dry, hot weather. For reasons that remain obscure, this fungus produces a complex of low molecular weight organic compounds containing, as shown in Figure 18, the unusual difurofuran moiety attached to a coumarin nucleus. These aflatoxins, for want of a more descriptive designation, may fulfill a necessary biochemical function in *Aspergillus* metabolism, but when consumed in human diets they derange metabolism, producing a highly mortal, fulminating liver disease with massive intestinal bleeding. The clinical picture of aflatoxicosis is one of acute liver injury—hepatitis. More recently, animal feeding studies have revealed that hepatoma—liver cancer—can be another manifestation of ingesting these naturally occurring but exquisitely fatal chemicals. The drought in the grain-growing areas of the United States in the dreadful summer of 1988 raised the real threat of *Aspergillus* contamination of corn (and hence of aflatoxicosis—contaminated by-products such as breakfast cereals and other corn-based processed foods. Indeed, the heat of baking these products does not detoxify these unique heat-stable chemicals. Thus the specter of widespread aflatoxin-containing grain immediately puts to test the nationwide grain inspection system, which for the most part must depend

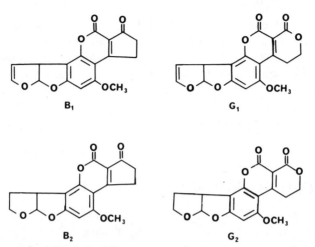

Figure 18. Aflatoxins commonly occurring as food and animal feed contaminants.

on the availability of experienced personnel, the number of samples taken for analysis, and the analytic procedures themselves. Fortunately, all the aflatoxins fluoresce in the presence of ultraviolet light. B_1 and B_2 emit a bright blue fluorescence while G_1 and G_2 radiate green light. This luminescence signals the presence of suspect kernels, which can then be subjected to further testing.

Although explosive aflatoxicosis has been a problem in India, an outbreak has never occurred in the United States. Whereas the prominence given the results of animal feeding studies has raised the threat of potential food-borne cancer risk, not a single case has ever been attributed to it. Nevertheless, sustained periods of hot, dry weather could overwhelm otherwise healthy plants, contaminating grain in the process and sorely trying our food safety practices.

GENE MANIPULATION AND ENHANCED PHOTOSYNTHETIC ACTIVITY

Photosynthesis is a remarkable process. The cascade of events, beginning with such simple ingredients as carbon dioxide and water and which results in the production of carbohydrates and proteins, is unmatched anywhere in the human world. Now, however, human ingenuity has begun a process of manipulating photosynthesis that will ultimately lead to a more efficient process; one that is faster and productive of larger plants. But just as important, these plants will have a resistance to such environmental chemicals as weed pesticides.

What a research team at Rutgers University's Waksman Institute has accomplished is nothing short of spectacular—transforming the genetic makeup of the tobacco plant, Nicotinia tabacum. Using a special "gun," they bombarded the chlorophyll-containing chloroplasts in the plant leaves with particles of tungsten coated with plasmid DNA and success-fully inserted foreign DNA into the chloroplasts of tobacco plants. The new DNA conveyed a new resistance to an antibiotic. This new resistance characteristic was incorporated into the plants genetic machinery and was passed on through seeds to their progeny, which now contain the trans-formed genes (19).

ROOT-NODULE ACTIVITY AND NITROGEN FIXATION

Peas, beans, and lentils, the legumes of this world, require nitrogen for their growth and have no trouble getting it. Rice, wheat, and corn, the major grains of the world also need nitrogen, but they have difficulty getting it. As shown in Table 1, Chapter 2, nitrogen is the atmosphere's most abundant gas, present at a level of approximately 80%. Still, nonle-guminous plants must obtain the nitrogen they need for protein manufac-ture by a more difficult maneuver.

Long ago, peas, beans, and lentils entered into a symbiotic and syner-gistic relationship with a remarkable group of soil microorganisms that reside in modules—some as small as gooseberries, others as large as grapes, attached to the plant roots. Here they spend their time extracting molecular nitrogen (N_2) from the atmosphere and converting it ("fixing" it) via nitrate to ammonia. This ensures the legumes of a ready supply of much needed nitrogen. In return, the bacteria obtain a supply of nutrients that they require. An outstanding example of a win/win situation. Rice, wheat, and corn have no such relationship and no nodules. If the soils in which their roots are sunk are deficient in nitrogen-containing com-pounds, they grow poorly unless inorganic nitrogen-containing fertilizers are added to their growing areas. For years, soil microbiologists and chem-ists have attempted to transmute rice, corn, and wheat into root-nodule types of plants—without success: until recently. Furthermore, recent field trials have shown that nitrogen fixation appears to be additionally stimu-lated by CO_2 enrichment. Anything that encourages plant growth will also increase nodule activity. Interestingly enough, laboratory experiments in-dicate that positive effects on CO_2 enrichment on nitrogen fixation con-tinue at least up to a level of 1000 ppm. If the almost 100 million tons of nitrogen-containing fertilizers used annually around the world could be dispensed with, the benefits would be manifold. Costs to farmers would be reduced, energy used in fertilizer production would be saved, and CO_2, an end-product of fertilizer manufacture, would be eliminated. It would be difficult to ask more of a scientific advance. Thus the recent report from the University of Sydney, Australia, of success in transforming wheat by im-planting root-nodule bacteria into its roots was heralded as a major agricultural advance. The research at Sydney suggests that it may indeed be possible to produce higher crop yields in nonleguminous plants without dependence on chemical fertilizers: that is, making wheat and rice independent of chemical fertilizers, permitting them to obtain their nitrogen needs in-house, as it were. The researchers at Sydney softened the root tissue with application of 2.4-D (dichlorophenoxyacetic acid) and then inserted the soil microbe *Azospirillium*, a nitrogen-fixing bacterium. In short order, they obtained substantial levels of nitrogenise, an enzyme whose presence indicates the occurrence of nitrogen fixation. Should this work withstand the rigors of further testing by other laboratories around the world, we would have further insurance of an undiminished food supply in the future. The key in these experiments would be the passage of nodule-forming ability to offspring via seeds.

GENETIC ENGINEERING

As noted earlier, genetic engineering techniques can be a source of new plant strains, but already available to farmers are both early and late maturing cultivars. Each of these can be substituted where and as required to compensate for changing seasonal temperatures. In Japan, for example, the use of late-maturing rice plants could well increase yields by 20–25%. In areas where growing seasons may be reduced by rising temperatures, switching to heat-tolerant varieties could offset any crop losses: just as drought-tolerant varieties can be used in those areas where drought is anticipated. Where one species of plant may be driven out of its current ecological niche by abruptly changing climate, another could take its place. *Zizania texana* is a case in point. *Zizania texana* is Texas wild rice, which is teetering on the brink of extinction. Its current habitat is limited to a single stream in central Texas, where it readily tolerates heat and dryness. These genetically expressed qualities could enable it to extend its range as well as that of wild rice now limited to the north central states. Plants such as *Zizania* offer insurance against the potential (or real) dislocations of a warmer, drier world. If the central states become hotter and drier, crops that do not find such climate stressful will be in demand. That plants can be developed with characteristics protective against existing weather and disease is evident in the investigations currently in progress at Birmingham University, England. Via intensive inbreeding, Harpal Pooni is developing a new breed of sunflower suitable for weather conditions in the United Kingdom. Because present sunflowers mature in the autumn, they are vulnerable to bad weather and disease. Pooni is attempting to produce a hybrid variety that matures earlier and is disease resistant. He expects to be able to achieve this goal in the next 3 years because of the special plant breeding techniques developed at Birmingham over the past 40 years. Clearly, our job is to search out plants that can thrive in what would ordinarily be excess moisture, heat, or dryness. There is little doubt that they are out there. We must also go back to the collections and locate specific seed stocks that can be prorogated and seedlings that can be planted in appropriate regions of the country and world. We are not without resources. Thus far we seem to be fixated on the effects of dislocations rather than on responses to change. Reasonable and available management practices can offset potentially damaging climatic stresses. Again, available and new management practices seem not to have been part of model bookkeeping and accounting equations. Obviously, sowing, planting, and harvesting along with the use of fertilizers and pesticides will need adjustment. This is what is entailed in change—different procedures. But this is not a crisis. It is a problem of reasonably manageable proportions, which, as history tells, is not new to farmers.

The evidence for response to change is growing. For example, scientists at the National Chemical Laboratory, Pune, India, devised a method that forced young bamboo plants to flower prematurely. Bamboo (*Bambusa arundinacea*) in the wild ordinarily takes 12–120 years to flower and produce seeds—one in their lifetime. The species used in the tests ordinarily require 30 years. With their enriched laboratory diet, they flowered after 3 weeks. And 2 weeks later healthy seeds appeared. With this new technique, botanists will be able to cross species to obtain faster growing plants and insert foreign genes for disease resistance and stronger wood. This is a breakthrough with significant implications for 25% of the world's population, considering that bamboo has such diverse uses as wood for homes, musical instruments, baskets, and scaffolding and pulp for paper. Contemporary botanical research fertilized with genetic engineering and tissue culture will be a source of many new plant varieties.

One of the most fascinating and far ranging reports for its implications for stabilizing food supplies in a warmer world is the recent work of DeKalb Plant Genetics of Groton, Connecticut, a subsidiary if DeKalb Industries—grain processors. The company has announced the successful production of fertile transgenic corn, transformed with a foreign gene that codes for an enzyme that endows corn with resistance to the herbicide Biolaphos. While herbicide resistance is not immediately germane to the warming problem, the development of fertility after insertion of a foreign gene is a major victory for genetic engineers because corn, a monocotyledon, had never before been so transformed. What this means is that plant breeders will have a technique for developing new varieties of corn rapidly and, most importantly, predictably, heretofore not possible with traditional breeding methods. In fact, breeders are looking at an abundance of riches because not one but half-a-dozen approaches are showing equal promise as companies, institutes, and university laboratories around the country and world race to obtain patents on new crop varieties. Transgenic corn and cereal grains or crops generally mean that these newly developed varieties have never before appeared on earth. They are artificially produced. While the motivation is often financial gain, because of the tremendous market, the benefits will accrue universally. And because of these crops develop fully in a single season, 5–10 years more will be more than adequate time to produce a number of varieties capable of high yields in warm and wet or warm and dry regions. Climate change, if and when it arrives, does not mean that we do not see to the new problems, the new needs. Human beings are like that—responding to new challenges and new directions. And lest in our fixity with the future we forget the past, it may be useful to consider Sylvan Wittwer's remarks on the subject. He informs us that "over the past 100 years, the high plains became the wheat belt during a moist period, then the Dust Bowl during a dry period." The essence is here. "Agriculture," he offers, "through migration and technology was able to adapt. . . . During the last century, one sees only upward trends attributed to new technology applied during both warm and cold and wet and dry periods" (10). That really is the essential point. We not only have persevered, we have prospered. And there is not reason to believe we will not continue to do so. Unless, of course, somewhere along the way we lose our will. Perhaps actual numbers will convey the degree of climate change to which farmers have responded in the recent past. Again, it is Wittwer recounting the historical record. "During the past century, American farmers coped with year-to-year standard deviations of 30 millimeters in eastern Kansas rainfall and two weeks in Minnesota growing seasons. From 1915 to 1945, Indiana farmers experienced a +0.1°C per year trend in temperatures and a total change of +2°C during that past century. . . . American agriculture already has demonstrated that it can adapt to a trend of +0.1°C per year, assuming no change in interannual fluctuations . . . Through the use of short season, early planted, single-cross maize hybrids, commercial production of corn in the United States has moved 500 miles further north during the past 50 years. The U.S. winter wheat zone could be moved 200 miles northward using a new level of winter hardiness now genetically available." That was 10 years ago, well before the current "crisis." Wittwer is not pontificating, not prophesying. He is a horticulturist who knows this country's agricultural history. He is refreshing to experience. Calamitous thinking is unnecessary.

EFFECT OF CLIMATE CHANGE ON LIVESTOCK

As a critical source of protein and other nutrients, livestock, cattle, sheep, goats, pigs, horses, and poultry must be considered in any global warming scenario. Adverse effects would likely present themselves either via available food supply or through physiological effects of heat balance. The former is obvious. Should foodstocks dwindle as a result of climate shifts, animals would be compromised. Similarly, widespread drought would exact major losses. If heat increases beyond an animals's capacity for internal cooling occurred, its appetite would fall, which in male animals would affect not only their weight and thereby meat yields but reproductive capacity as well. Obviously unmanageable temperature increases could have far-reaching deleterious effects. Additionally, warmer temperatures will require augmenting air-conditioning facilities to prevent decreases in egg-laying in poultry and to avert milk production decreases and weight loss in cattle. Added costs of energy use would be passed on to consumers in the form of higher prices for these items, unless energy costs could be maintained or reduced by more energy-efficient management.

Screwworm Infestation

Insect pests of cattle could exceed their current ranges if regional weather patterns shift. The screwworm fly and horn fly are illustrative models. The screwworm fly (*Cochliomyia hominovorax*), a scourge of cattle, does not ordinarily overwinter north of Florida and the extreme southern part of Texas, but it does occur much farther north in mild winters. Thus warmer southern winters could see the screwworm break out of its normally temperature-restricting boundary. Adults are strong fliers and can cover 25–35 miles per week.

Screwworms are about three times as large as the common horsefly and attack cattle unmercifully. The adult female screwworm fly lays a mass of 200–300 eggs in the wounds of cattle, sheep, goats, or horses. Any wound, accidental or surgical, a tick bite, or the navel of a newborn animal can be an egg-laying site. From 12 to 24 hours later, maggots (larvae) hatch and begin to feed in the wound site. A hundred mature maggots can produce a wound about 1 inch in diameter and 1 inch deep. The feeding causes an additional bloody discharge whose odor attracts more flies for additional egg-laying. Death of the animal is inevitable unless it is found and treated. Even though treatment can save the animal, the hide will have holes and blemishes that reduce its value, and the irritation caused to cattle in particular reduces milk production.

Mature maggots leave the wound, drop to the ground, enter the soil, and transform into the pupae stage. Flies emerge from the pupae in 8–30 days depending on the air temperature. The adult flies mate a few days after emerging. And herein hangs the means of reducing their predations whatever the temperature.

The sterile-male technique was pioneered on the screwworm fly. In this eradication program, screwworm flies are sterilized by gamma radiation from a cobalt-60 source. In the sterilization procedure, thousands to millions of fly eggs are collected on screens or in troughs. The eggs are placed on an artificial food until the larvae emerge and are reared through pupation. The pupae, in plastic containers, are placed in a gamma cell to be exposed to radiation from a cobalt-60 source. The radiation is sufficient to sterilize them without reducing their activity. After sterilization the pupae are placed in paper bags, in which they develop into adult flies. These bags of flies are placed in dispersing tubes aboard an airplane. A tube can handle one bag every 2–3 seconds and thus is capable of dispersing millions of flies over a wide area. Air, streaming through the dispersing tube, whips the bag out of the tube. As it leaves the tube, the bag is slashed by a hinged knife; the bag sticks on a hook for a moment and the flies are scattered.

The sterile-male technique has three important advantages over the conventional application of chemical pesticides. In the first place, it is highly selective, involving only a specific insect while leaving all other forms (insects, worms, birds, plants, etc.) undisturbed. In addition, the target species cannot acquire immunity to sterile mating, as it too often does to chemicals. Finally, chemical agents often become less efficient as the population against which they are being used declines. As a consequence, the few survivors can begin to rebuild their decimated ranks. On the other hand, the sterile-male procedure becomes even more efficient as the population dwindles. Actually, the sterile-male technique should be called the sterile insect technique (SIT) because both sexes are sterilized and released together as the cost of separating the sexes in mass-reared chambers is prohibitive.

An instructive case in point currently presented itself. Libya is the first country outside the Americas to be afflicted by the screwworm where it was discovered in 1988 having been imported on infested livestock. By 1989 it had spread rapidly; in 1989 there were fewer than 2000 known cases. By the latter half of 1990 there were over 12,000. The fear is that the North African coast, southern Europe, and countries south of the Sahara will also become infested. The climate is perfect for the flies to develop a habitat. But the plan now in effect calls for their eradication, and the procedure chosen is the SIT. Currently, some 7 million flies are being released from aircraft weekly. But this is expected to increase 100 million by July 1991 when natural fly population are at their highest. Theoretically, eradication should result in total removal. In fact, it hasn't. As long as feral pigs, boars, and other such types exist, the flies cannot be totally eradicated. But they can be kept at extremely low levels; low enough to prevent managed cattle from being infested. Should global warming bring a shift in regional weather patterns such that the screwworm can expand its boundaries, there would be means readily available to prevent its numbers from increasing. And recall that regional weather patterns will not shift overnight: it will take years to decades; enough time to respond.

Hematobia Irritans

The horn fly is a small black fly half as large as a housefly but it is responsible for some $600–700 million losses to the beef and dairy cattle industry. Most of their adult life is spent on and biting animals. It lays its eggs on fresh manure, wherein the larvae feed and mature. Cattle lose energy and weight when constantly fighting off attacks of horn flies. Animals protected from the flies can gain half-a-pound a day more than unprotected animals. Thus higher temperatures, which could extend the flies breeding season and thus the length of the biting season are of concern. On the other hand, increased temperatures in the southern states could well become intolerable for the flies, thereby reducing their activity. In northern areas, effective control measures can also severely limit their activity.

EFFECT OF CLIMATE CHANGE ON INSECT PESTS OF CROPS

As for insect pests of crops, a doubling of CO_2 should not affect them directly. Of course, CO_2 will affect plants and in so doing may indirectly discommode insects that feed on them. For example, if more carbohydrate is produced as a consequence of increased photosynthesis, more sugar would be present and thus attract more insects, especially bees and other pollinators. On the debit side, of the few feeding studies that have been undertaken and reported, one finds that with increased CO_2, soybean leaves had a greater carbon to nitrogen ratio, which when fed to soybean looper larvae required extended feeding periods to extract a level of nutrients equal to leaves of lower C:N ratios. This type of study suggests that insect predation may increase. On the other hand, studies of cotton plants grown under enriched CO_2 do not show increased sugar content and no increased insect feeding activity. Given the paucity of data thus far available, interpretations and conclusions are limited. Clearly, some plants will be attacked at an increased rate, while others will not.

ADVERSE HUMAN HEALTH EFFECTS OF OZONE DEPLETION

The consequences of ozone depletion proceed from the belief that, with an ozone layer in tatters, penetration of ultraviolet radiation would increase markedly at sea level with subsequent adverse human health effects. But this would in great measure depend on the amount of UV penetrating and where the penetration occurred. It is one thing to assume equal and severe penetration and thus equal risk to the entire world population and extrapolate increased incidence of skin cancers based on penetration during the spring in areas contiguous to the south polar region, for example, where ozone depletion has been significant, and the high northern latitudes in winter when depletion is low, but greatest there. The latter is a more demanding undertaking because it requires distinguishing potentially at risk target populations within the depleted zones. However, in addition to calculating numbers of potentially exposed individuals and by extension estimated numbers of additional illnesses to be expected, it would be useful to determine the levels of UV penetrating to the bottom of the troposphere and to determine if the levels are in fact increasing. Are UV levels at the earth's surface directly related to density of the ozone layer?

Currently, the U.S. EPA maintains that with ongoing ozone depletion "grave consequences" for both human health and the environment would ensue. Accordingly, in 1987 the EPA developed a risk assessment that predicted increases in skin cancer, cataracts, and suppression of the human immune system with the attendant risk of infectious disease that that would entail. Based on their calculations, shown in Table 4, the increase in UV-B resulting from a 50% reduction in the ozone layer stretching across the entire country would directly cause millions of additional cases of skin cancer among the Caucasian population. This assessment was limited to the most susceptible group in the population. Dark skinned individuals with high levels of the pigment melanin in their skin are well protected against UV radiation. In effect, the melanin acts as a protective ozone layer. As Table 4.4 shows, substantial differences in numbers of adverse events are estimated with and without a 50% reduction in emissions of chlorine-containing halogenated compounds. With their unabated use, the EPA assessment calls for 178 million additional nonmelanoma skin cancers. If by the year 2080 the population of the United States is approximately 300 million, then 178 million skin cancers (basel and squamous cell carcinomas) would mean that close to 60% of the population would be affected; an unprecedented number for cancer of any type but perhaps reflective of the severity of cellular damage inflicted by excessive UV-B radiation. By comparison, a 50% reduction in the use of CFCs and a freeze on Halons would yield approximately 5 million additional cases. In this instance the loss of ozone is less than 2%. Thus in their assessment scheme, even a small depletion yields fairly large numbers of new cases, suggesting the deleterious potency of UV-B. As for melanoma, a highly malignant carcinoma afflicting substantial numbers in the best of times, close to 1 million cases are projected without control of emissions and less than 50,000 with control. Obviously, if these numbers are credible, even a small reduction in

Table 4.4
Potential Additional Skin Cancer and Cataract Cases in the United
States for People Born by the Year 2075

Figures in thousands

Scenario	Ozone depletion by 2075 (percent)	Non-melanoma cancer (whites only)		Melanoma cancer (whites only)		Cataract cases
		Cases	Deaths	Cases	Deaths	
No controls	50	177,998	3,529	893	211	20
CFC 50% cut & halon freeze	1.9	5,104	81	46	11	876

Source: U.S. General Accounting Office, February 1989.

ozone layer thickness is to be eschewed. Furthermore, if the ozone layer is thinned as a consequence of CFC catalysis, the greatest increases in UV penetration are expected to occur in the spectral region between 290 and 310 nanometers (nm), the region of UV-C and UV-B; the least increase in UV penetration would be at 320 nm and above the region of UV-A. As shown in Figure 19, the ozone layer effectively filters our energetic radiation below 320 nm. From the figure it is evident that even with large ozone depletions little UV-C should penetrate to the earth's surface. Of additional pertinence are the studies of Doniger and co-workers at the National Cancer Institute, who investigated the lethal effects of UV light on animal cells. We see in Figure 20 that certain wavelengths have greater biological effectiveness than others. For wavelengths of 297, 303, and 313+ nm, it is seen that far greater levels of radiation (panel c) are needed to induce pyrimidine dimers in DNA. Pyrimidine dimers are the major UV radiation-induced product responsible for such untoward biological effects as mutations and cell death. As shown in the figure, covalent bonding of two adjacent pyrimidine molecules (open squares) is facilitated compared to their effectiveness in inducing neoplastic cellular transformation (black squares). The solid and dashed lines are the linear regression lines for the transformation and pyrimidine dimer data. The ill effects of UV on DNA fall precipitously as wavelength changes from 290 nm to 320 nm; from 290 to 320 nm the decrease is approximately six orders of magnitude.

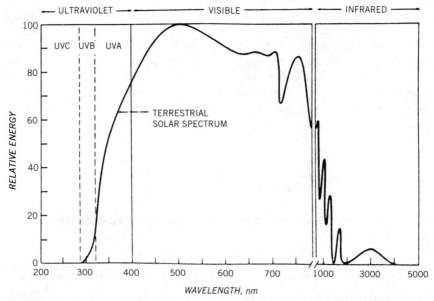

Figure 19. Spectrum of electromagnetic radiation that reaches the earth's surface from the sun. Wavelengths shorter than approximately 290 nm are absorbed by ozone in the stratosphere. (Courtesy of the U.S. EPA.)

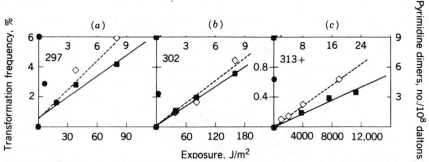

Figure 20. Effectiveness of 297, 302, and 313+ nm UV radiation at producing pyrimidine dimers (□) and transformation (■). (Courtesy of the U.S. EPA.)

With this as a point of departure, let us look at current UV-related adverse human health effects in greater depth. In addition to melanotic and nonmelanotic tumors, solar radiation can induce ocular abnormalities including cataracts, a clouding of the lens of the eye leading to loss of vision, as well as changes in the retina, a membranous layer at the back of the eyeball. Medical evidence also suggests that solar radiation can suppress immune system responses.

When excessive doses of UV radiation, that is, biologically effective doses, strike epidermal cells, it is the DNA of the cell nuclei that are in fact the target molecules. However, because the body is not a passive recipient of untoward environmental stimuli, its natural defenses respond, and the process of repair ensues.

As a result of energy absorption, a lesion occurs; two nearby DNA units form an abnormal chemical bond, a dimer, a likely response to the formation of noisome organic ring compounds within the nuclei. But these are not necessarily lethal lesions. Outwardly observable, the process appears as erythema (sunburn) with consequent blistering, peeling, and new skin formation—all part of the process of DNA repair, in which enzymes attach to the damaged site and clip out molecules on either side of the damage. The entire segment is then removed, allowing a second set of repair enzymes to fill the void and form a new batch of healthy cells, DNA repair is the rule; deranged cellular proliferation, carcinogenesis, and lack of repair are the exceptions.

The interaction of sunlight with the skin is a manifold process, involving the transfer of solar radiation to a spectrum of molecules within the epidermis—the uppermost layer of skin—and the subsequent cellular response to this energy transfer. Some 95% of the incident radiation penetrates the skin; the remaining 5% is reflected by the stratum corneum, the thin layer of cells at the top of the epidermis. If past measurements are accurate, from 1 to 20% of sea level UV penetrates white skinned individuals to the depths of the basal layer of cells, and the 1–20% reflects the varying thicknesses of the epidermal layer and stratum corneum in different individuals as well as inherent levels of UV radiation-absorbing chromophobe molecules. Among the wide variety of chromophobes is melanin, which absorbs most of the UV in the skin. As noted previously, the skin is not simply a passive recipient of UV radiation. As exposure increases, melanocytes produces more melanin at increasing rates of exposure. Tanning begins to occur 10–12 hours later and thickening or keratinization of the skin occurs simultaneously, inducing a protective effect. And there is another major benefit to be derived from UV exposure. Fat-soluable vitamin D_3—cholecalciferol—is synthesized photochemically in the skin by UV irradiation of 7-dehydrocholesterol. Following hydroxylations in both the liver and kidney, cholecalciferol is converted to 1,25-dihydroxy cholecalciferol, the most active form of the vitamin. Thus vitamin D, essential for protection against the anomalous bone conditions of rickets in children and osteomalacia in adults, is unique among vitamins because the body is not solely dependent on external sources of it—that is, the food supply—to meet its needs. On the one hand, vitamin D is absorbed directly from the skin into the bloodstream and additionally is ingested with food and absorbed into the bloodstream with fat. Thus it is clear that UV radiation is essential for the integrity of the human system. Nevertheless, epidemiological evidence is clear that skin cancer is the most common form of cancer among Caucasians and is strongly related to UV exposure. In fact, a melanoma epidemic has been building in the United States since the 1950s, well before the ozone layer became a concern. Evidence suggests this to be the consequence of our national love affair with tanning and sun worshiping. Exposure to the carcinogenic action of UV radiation is much greater and begins at an earlier age than exposure to any other widespread environmental carcinogen. In 1990, some 27,600 people were expected to develop malignant melanoma, which arises in areas of the skin, mucous membranes, eye, and central nervous system, where pigment cells are found. Melanoma is a tumor of the melanocytes—the cells that produce skin pigmentation—and on the skin lesions can range in color from white or red to blue, brown, or black. The discolored area is often raised with notched edges that bleed easily. In the 1930s one person in 500 was expected to develop melanoma. By 1990 the figure had climbed to 1 in 120, with the greatest rate increases among women under age 40. In fact, it is now the leading cause of cancer among women aged 25–29. Although melanoma represents only 3% of cutaneous neoplasms, it is responsible for over 60% of all skin cancer deaths. Clearly, it is extremely malignant. Blistering sunburns during childhood and adolescence appear to trigger the malignant transformation, which then takes 15–30 years to express itself. Here we see the relationship between duration and intensity of exposure. Sunburn is the major risk factor but who gets the disease is a constitutional matter involving heredity and individual susceptibility. Some 10% of all cases run in families. Other risk factors are fair skin, blond or red hair, and marked freckling of the upper back. Caucasians are in fact at 60 times the risk of skin cancer compared to blacks. The incidence rate for whites is 242/100,000 compared to 4/100,000 for blacks. Data from the American Cancer Society indicate that living closer to the equator—which would increase exposure—is yet another risk factor. For example, the melanoma rate was 6/100,000 for people in Detroit, 11/100,000 for people in Atlanta, and 27/100,000 for people in Tucson. Figure 21 highlights this latitudinal difference. Clearly, incidence increases as one moves from Seattle to New Orleans. This stiking latitudinal difference is further highlighted by data from England and Australia. British men have a rate of 1.9/100,000 compared to 40/100,000 in Queensland. Quite clearly there are important hemispheric differences that will continue should holes open in the ozone layer.

In January and February 1989, ozone levels in the region of the Arctic circle were found to be 25% lower than the region below the Arctic. NOAAs high altitude aircraft flying from Stavanger, Norway toward the north pole from January 3 through February 10, at altitudes of 9–12 miles, found that, as in the south polar region, the area of maximum depletion also occurred with a vortex of cold, high winds that form over the pole in winter. Column ozone readings (DU's) were found to be from 12 to 35% lower within the vortex compared to outside. This translated to a final value of some 6% reduction in column ozone, the total amount over a given

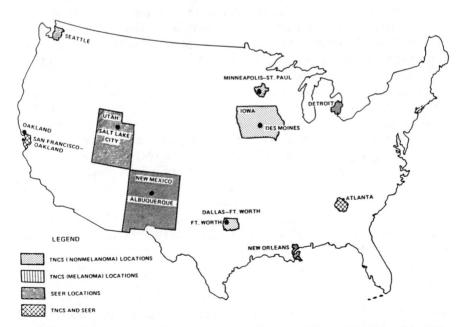

Figure 21. Skin cancer measurement locations in the United States. (Courtesy of the U.S. EPA.)

spot on earth. Over the south polar area reductions of as much as 50–70% in total column ozone have been recorded. However, according to Michael Proffet of NOAA's Aeronomy Laboratory, these levels of ozone reduction should not be expected over the Arctic because the meteorological conditions over the Arctic produce a vortex of far shorter duration, which is directly related to ozone reduction. Thus the ozone losses in the Arctic are in no sense a hole as has occurred over the Antarctic. For those people, the Swedes, the Danes, the Scots, living in the higher northern latitudes, this could be a source of some comfort.

According to projections of the American Cancer Society, over 600,000 cases of the more common nonmelanotic basal cell and squamous cell skin cancers were projected to occur in 1990. These usually occur on the face and hands, rarely invade deep tissues, and are highly curable. Basal cell tumors are generally slow growing and most often benign. They appear as small, shiny firm nodules with a centrally expressed area, but they can also be difficult to differentiate from psoriasis, a localized dermatitis. They are commonly found on the back of the hands and face, often ulcerate and bleed easily, and appear never to heal. Squamous cell tumors, the second most common form of skin cancer, are similarly slow growing and mostly benign and can be seen as red papules with white to yellow scaly or crusted surfaces, which can also ulcerate and bleed easily. Basal and squamous cell tumors rarely metastasize, but on rare occasions can become fatal. Thus professional consultation is necessary for tumor formation to be separated from other similar appearing but nontumorous dermatologic discolorations. The fact of an estimated 600,000 expected new cases in the United States in 1990 is an important statistic because it is from such expected numbers in periods of normal UV-B levels that projections to atypical periods are made.

Americans could easily halve both their melanoma and nonmelanoma rates by judicious application of level-15 sunscreen creams and avoidance of sun worship between the hours of 10 a.m. and 2 p.m. when the summer sun is strongest in the northern hemisphere. A bit more cover-up would also go a long way toward self-protection. Certainly these would be appropriate activities for preteen children. Projections of increased cancer rates are made without factoring in preventive measures, which in the case of solar radiation-induced cancers can be substantial.

Recently, two young people with malignant melanoma, a 42-year-old man and a 29-year-old woman, had a gene for an anticancer toxin implanted in their bodies. They were the first people to receive gene therapy in an attempt to reverse an otherwise incurable tumor. The experimental procedure undertaken at the NIH Clinical Center, Bethesda, Maryland, seeks to increase the cancer fighting ability of the patient's own tumor cells by providing TNF—tumor necrosis factor—at the site of the lesions. With the newly implanted gene, TNF in large amounts will be brought directly to the tumors, concentrating their enzymatic action on the rapidly growing cells and eroding them. The idea continues the long tradition of searching for a "magic bullet" that would destroy malignant cells without affecting healthy ones. Should this pioneering trial prove successful and the melanoma go into remission, then a second trial will undoubtedly be required to assure that the first was not a chance occurrence. If the results are nonequivocal, genetically engineered enzyme therapy will become widely used. This supposes that the problems surrounding the procedure of loading white cells with appropriate genes for producing tumor necrosis factor are surmounted. If all goes as hoped, by the year 2000, melanoma could be relegated to the scrap heap of yesterday's diseases. Another victory for mankind—by mankind. At the right time.

OCULAR DAMAGE

Medically speaking, cataracts are a degenerative opacity of the lens often characterized by progressive loss of vision. In addition to exposure to sunlight, cataracts are known to be the result of advancing age (senile degeneration), x-rays, diabetes, and a variety of systemic medications. There is little doubt that intense exposure to UV-B can injure the structures at the front of the eye, but whether it produces cataracts or macular degeneration remains to be demonstrated. Circumstantial evidence suggests that it does.

The anatomical structures of the eye impose a series of reasonably good barriers to the penetration of UV radiation. It is the cornea that absorbs the most energetic component of UV-B, and the cornea has built-in damage control and repair mechanisms. But as it well established, intense UV radiation refected off sand, water, and snow can cause a transient, but painful, corneal burn commonly known as snow-blindness. Photokeratitis is the medical term. It is the lens that is responsible for focusing light on the retina and it is the lens that absorbs the brunt of the UV radiation and passes only visible light to the retina. Whether UV radiation induces cumulative damage to the lens remains an open question. Laboratory studies have, shown however, that UV light can produce structural changes in lens protein. For the lens to remain transparent, its protein must remain properly organized. UV radiation has been shown to disrupt its organization, leading to opacity. Of 600,000 or so people undergoing cataract surgery each year, the greatest number of cases appear to be the result of old age. But this can change. Cataracts are also known to vary in incidence by latitude; consequently, should a hole in the ozone layer open at the south pole for a sustained period, and if it is sufficiently wide, increases in ocular damage could be seen in Australia and New Zealand. Here again, damage can be averted and reduced by the use of sunglasses expressly made to filter out UV light. As shown in Figure 22, all sunglasses do not do this. To be sure, a statement to this effect must accompany the

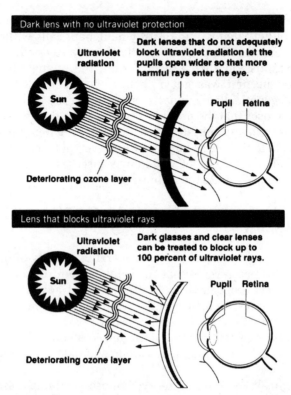

Figure 22. Eye safety: illusion and reality. Special treatment of lenses can protect against UV radiation. (Reproduced with permission from *The New York Times*.)

glasses. And for optimum protection they should be worn as close to the eyes as possible.

Additional protection is available for anyone willing to wear glasses. Lenses can now be coated with a chemical that absorbs UV radiation in the 290–400 nm range, dissipating the potentially hazardous energy before it reaches the surface of the eye. Inland Diamond Products of Madison Heights, Michigan, developed an organic compound that has been used for over a decade in the textile and plastics industries to prevent sunlight-induced product degradation. Apparently, the UV radiation is reflected within the coating itself, expending its energy between the double-bonded carbon atoms that form the backbone of the molecule. More recently, their compound was adopted as a coating for lenses affording protection against UV damage to the eyes of forest rangers who spend extended periods outdoors. Now the coating is becoming available to the public through optometric outlets. While fees can be expected to vary around the country, $30 seems to be a nationwide average. As the market responds to increased demand, the price should drop. Both glass and plastic lenses can be coated, and this added protection is not limited to those with prescription lenses.

IMMUNOSUPPRESSION

Photoimmunology is a relatively new area of scientific investigation. It refers to the potential and actual effects of solar radiation on the immune response—that is, the effect of light of various wavelengths on the ability of the body's natural defense mechanism to repond appropriately when needed. It is the body's in-depth defense mechanism that helps us remain disease free. Disease free is of course the norm. Illness is the exceptional state and implies an overriding of the body's defenses.

In the 1970s several published studies suggested that tumors transplanted from cancerous animals to healthy animals were more successful, that is, continued to grow, when the noncancerous animals were exposed to UV radiation prior to transplantation. Other studies suggested that UV light may suppress the immune system, permitting the activation of herpes virus cold sores. But this has also been attributed to nutritional deprivation and stress. It has also been hypothesized that UV radiation could have a deleterious effect on the immune response to those infectious diseases that enter by way of the skin, especially if the initial immune response to the microbe occurs in the skin. But until such time as rigorous studies are performed, this conjecture must remain an untested concept.

Immunosuppression may reasonably be expected to exacerbate risk to those individuals currently immunosuppressed, such as transplant patients or those with concurrent immunodeficiency diseases. But this also remains to be investigated. Another major gap in our knowledge is a lack of data from which a dose–response relationship can be constructed. Anecdotal information suggests that laboratory animals require less UV radiation to induce immunosuppressive reactions than that needed to induce carcinogenesis. If this were in fact found to be the case, and the data were transferable to the human system, it would mean that doses of UV even lower than those needed to induce sunburn could be a problem. Until such studies are undertaken and their results analyzed, the suggestion remains just that. Immunosuppression is an area of unsubstantiated research but tantalizing portents. Thus, until such time as the requisite controlled studies are undertaken and verified, immunosuppression effects currently attributed to UV radiation must remain questionable and unproven.

MEASUREMENT OF BIOLOGICALLY EFFECTIVE UV RADIATION

The damaging effects of UV penetration are dependent on the amount of radiation reaching sea level. Several as yet inconclusive studies suggest that a 1% decrease in stratospheric ozone could increase the amount of UV-B passing through the troposphere by 2%. Considering that substantial evidence now supports CFC photocatalyzed ozone depletion since 1969, at times as high as 50% in the south polar area, what evidence can be mustered to show the actual presence of increased UV-B levels at the surface of the earth?

In 1974 a collaborative effort between the National Cancer Institute, the National Oceanographic and Atmospheric Agency, and Temple University's Health Science Center was undertaken to measure nonionizing solar radiation at ground level—specifically, UV-B at 290–330 nm. Using Robertson–Berger (R-B) photosensitive meters, measurements were made at eight locations, usually airports, across the United States. In 1988, with 12 years of data in hand, they reported that, over that period, "there are no positive trends in annual R-B counts." And they went on to note that the study suggested that "the role of physical and meteorological factors in the troposphere may be greater than expected, and that there may be prevailing conditions that diffuse solar energy and thus reduce the amount of UVB radiation reaching the earth's surface." Their surprising data notwithstanding, they nevertheless maintained that increased exposure to UV-B should produce large increases in nonmelanoma skin cancers and increased melanomas in areas near the polar region (20). Two years later, Blumthaler and Ambach of the Institute of Medical Physics, University of Innsbruck, reported that their measurements, obtained some 11,000 feet above sea level, indicated increases of UV-B on the order of 1% per year since 1981. Their measurements were restricted to cloudless days. Furthermore, they reported particularly high UV-B values in the spring of 1983, corresponding to reduced ozone after the eruption of the El Chichon volcano in Mexico, which indicates that ozone can be depleted by chemicals other than the CFCs (21). It is entirely possible that both of these reported measurements are misleading and that R-B meters are inadequate instruments for the purpose. It may be that a global network of ozone-measuring instruments should be established in remote, unpolluted areas to obtain long-term trends similar to the CO_2 monitoring stations. In addition, data from orbiting satellites with direct reading instruments could readily obtain measurements in the troposphere. Apparently, plans do call for TOMS type of apparatus to be placed aboard Russian, Japanese, and possibly American satellites to obtain UV–ozone data during the period 1992–1994. In the meantime, and until reports of more substantial penetration are forthcoming, a brief discussion of personal protection is warranted. If the Arctic region—because of its local meteorological conditions—cannot be expected to undergo the level and extent of ozone reduction encountered in the Antarctic, it should not be expected elsewhere, unless of course an entirely different set of conditions can destroy ozone. Unless and until such a system is uncovered, ozone depletion for the great majority of the people of the world should not be the problem initially envisioned. For most people, malignant and benign alterations of dermatologic cells will occur as a consequence of overindulgence in sunbathing. Until prudence takes hold, the incidence of both basal and squamous cell carcinomas, as well as malignant melanoma, will continue its execrable rise. At this juncture we enter the realm of potential versus actual exposure dose and potentially biologically effective dose versus actual biologically effective dose. The differences are substantial. For example, the total amount of energy available to an individual is one thing but the actual amount falling on the skin surface depends on the amount and type of clothing worn and/or the amount of time outdoors, as well as personal activity such as positioning oneself relative to the direct rays of the sun—between 10 a.m. and 2 p.m., or after 2 p.m., when the angle of the sun is less than 45°. Similarly, the biologically effective dose is the delivered dose to the skin and is directly dependent on the actual time and type of exposure and state of the dermatologic pigmentation, along with a host of natural environmental factors such as locations (latitude), season, time of day, cloud cover, and surface reflectivity—that is, beach (sand and water) versus forest or mountains. Additional factors that affect effective dose are artificial or manufactured items such as cosmetics, suntan lotions, and sunscreens and sun reflectors used to enhance the amount of radiation striking the skin. All these factors must enter the equation for predicting deleterious effects.

To this can be added the year-round level of tanning. Dark skin and tanned white skin allow less of a unit dose to penetrate the basal layer of the skin than lightskin. Should tanning occur relatively slowly during the spring and early summer before the sun becomes strong and more UV-B is available, compared to the more intense exposure that can occur in June, July, and August, the actual dose is far different, because early tanning becomes protective, allowing less radiation to reach the basal layer. In addition, epidermal thickness, especially the thickness of the stratum corneum, can effectively block radiation penetration. The stratum corneum (Figure 23) contains the protein keratin, which is an absorber of UV radiation. The more keratin present, the greater the protection afforded. Thus some people are constitutionally better able to withstand the potentially ill effects of solar radiation, while others must protect themselves externally. In the long run, it does become a personal and individual matter. Some will, of course, overdo their exposure to the sun to their detriment. At this point a word about sun blocking or sunscreen lotions and creams is in order.

Currently, the numbers of protective levels 2 through 15 can be misleading. There is no national or international standard to which each manufacturer must adhere. Thus the degree of protection afforded against UV-B that a specific sun block indicates it offers may be meaningless if different brands or manufacturers use different and incompatible analytic procedures to obtain their protective level designations. Until industry-wide standards are adopted and cover both domestic and foreign products, their protective effects remain equivocal.

In addition to protective lenses and sunscreen creams, worries about the harmful effects of sunbathing could be reduced by a small, inexpensive

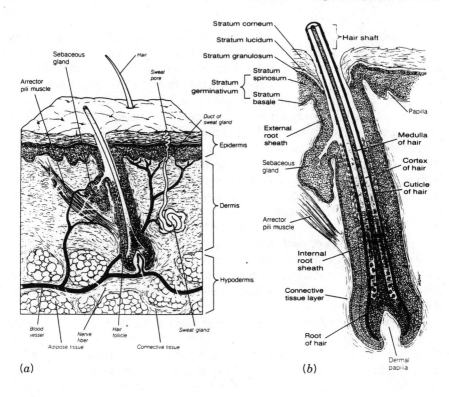

(a)

(b)

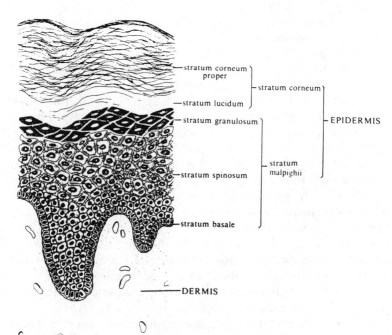

(c)

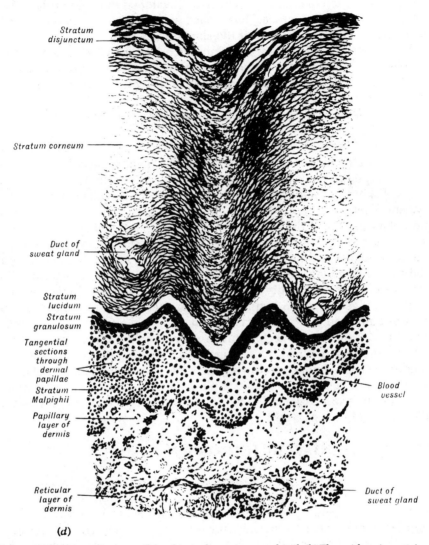

Stratum
disjunctum

Stratum corneum

Duct of
sweat gland

Stratum
lucidum
Stratum
granulosum
Tangential
sections
through
dermal
papillae
Stratum
Malpighii
Papillary
layer of
dermis
Reticular
layer of
dermis

Blood
vessel

Duct of
sweat gland

(d)

Figure 23. (a) The architecture of the skin is shown in some detail. (b) The epidermis contains the stratum corneum (Sc). (c, d) The stratum corneum (cornu = horn) is the most superficial layer of the epidermis, but it can have considerable depth. Since the cytoplasm of cells in the Sc has been replaced by a fibrous protein, keratin, these dead cells are said to be cornified or horny. The outermost layers of the Sc are constantly being lost as a result of abrasion, but they are also constantly being replaced by cells from the deeper layers of the epidermis.

hand-held suntan measuring instrument developed in England. The unit, which weighs only 2 ounces, has a narrow cylindrical probe that illuminates 1 centimeter (0.4 inch) of the skin surface with red light. The skin reflects a portion of this light back into the meter, which translates it onto an electrical signal that produces a voltage directly proportional to the light received. The signal is converted to an illuminated display, indicating the amount of exposure and advice on dangerous levels. Inquiries about this device, which as yet remains unnamed, can be addressed to the Design Council, London.

Earlier it was noted that in 1987 the U.S. EPA performed a risk assessment of the human consequences of a depleted ozone layer and found that increased penetration of UV-B radiation would induce millions of cases of skin cancer and cataracts and would suppress the body's immune system. Their most recent assessment follows hard on the recent observation of a thinning of the ozone layer above the continental United States. Their prediction of new cases of skin cancer over the coming 50 years is of the order of 12 million (±25%) and 200,000 accompanying deaths. But there is an important caveat here. Their numbers proceed from the belief that people will not modify their behavior—behavior developed over the past 15–20 years, which maintains that greater nudity in the sun on the part of both men and women is both fashionable and appropriate: that, plus the idea that longer periods in the early afternoon sun are preferable to late afternoon sun, when the rays are longer and less severe. If people do modify their behavior by following the suggestions noted here, especially avoiding sun between noon and 2 p.m., and by covering up a bit more, particularly arms, legs, and face, the 12 million predicted cases would decline precipitously.

UV AND THE ANTARCTIC FOOD CHAIN

The decrease in atmospheric ozone level over Antarctica, the Arctic, or anywhere else implies increased levels of UV radiation recorded at the earth's surface. Reductions in total column ozone are a concern for their implied ill effects to a broad spectrum of life forms. But there is a question as to whether there is a direct relationship between these measurements and actual ill effects. In fact, a number of researchers have shown that leaf structure and protective pigment content are key elements in protection between the studies of plant physiologists and climatologists/atmospheric scientists. The climatologist/atmospheric scientists are concerned with measuring levels of ozone and UV, and to their credit they do raise the question of how changing levels, especially decreases in ozone and corresponding increases in UV, may adversely affect living things. That's as it should be. Too often, however, simply raising the question takes on a life of its own and is translated as decreasing the ozone layer will destroy the Antarctic food chain with all its far-reaching ramifications.

Plant physiologists have studied the relationship and have found that plants possess adaptive mechanisms that protect them against a variety of adverse environmental stimuli. For example, it has been well documented that plant leaves can effectively filter out the short wavelength UV-B (280–320 nm), preventing them from reaching the chloroplasts, while permitting minimal attenuation of the 400–700 nm band, which includes the necessary photosynthetic light energy (22). The recognized existence of this selective light filtration system makes it unlikely that in their natural habitats plants will be damaged by UV-B radiation. The presence of this protective mechanism is of twofold importance: under conditions of reduced atmospheric ozone permitting penetration of increased UV-B, and in instances where plant species migrate to areas of naturally higher UV-B radiation. It is evident that plants have the ability to modify the optical properties of their leaves, achieving for themselves the ability to tolerate intense UV-B radiation (21, 22). It has also been well demonstrated under glass in laboratories as well as in the field that plants respond to increases in UV-B irradiation by increasing tissue levels of bioflavinoids. And it has been demonstrated that induction of flavinoid synthesis continues with continued UV-B stimulation. Again, photosynthesis is not impaired by increasing levels of UV-B. Plants grown naturally have been shown to be much less sensitive to increased UV-B radiation compared with similar plants grown under hothouse conditions. Additionally, the naturally grown plants do not show the reductions in per acre yield shown by plants grown in growth chambers.

Most recently, Kenneth R. Markham and co-workers of New Zealand's Department of Scientific and Industrial Research analyzed the flavinoid content of herbarium moss species collected in the Ross Sea Area of Antarctica between 1957 and 1989. Their flavinoid record is directly related to the levels of atmospheric ozone. Consequently, a continuous record of ozone over a 30 year period appears to have become available (24). Although continuous records of ozone levels in Antarctica over the same periods do exist, they relate only to two sites. As the New Zealand researchers note, "data for regions and periods other than these are generally accepted as being unobtainable now." Thus plants collected at a point in time, over an extended period may well be a "barometer" of ozone levels to which they were exposed. Figure 24 shows the inverse or mirror image relationship between flavinoid content and ozone level. High ozone levels mean that little UV-B radiation penetrates through to strike plants at the earth's surface, hence the protective flavinoid pigment levels are low; protection is not necessary. When ozone is low, UV penetrates and plants require protection and do so by producing flavinoid in high concentration. Markham and co-workers point out that while the trend levels show the now well documented ozone depletion of the mid- to late 1980s, they also show a surprising reduction in ozone levels in the mid-1960s. This 1960 decline has as yet received scant attention. However, the data do suggest

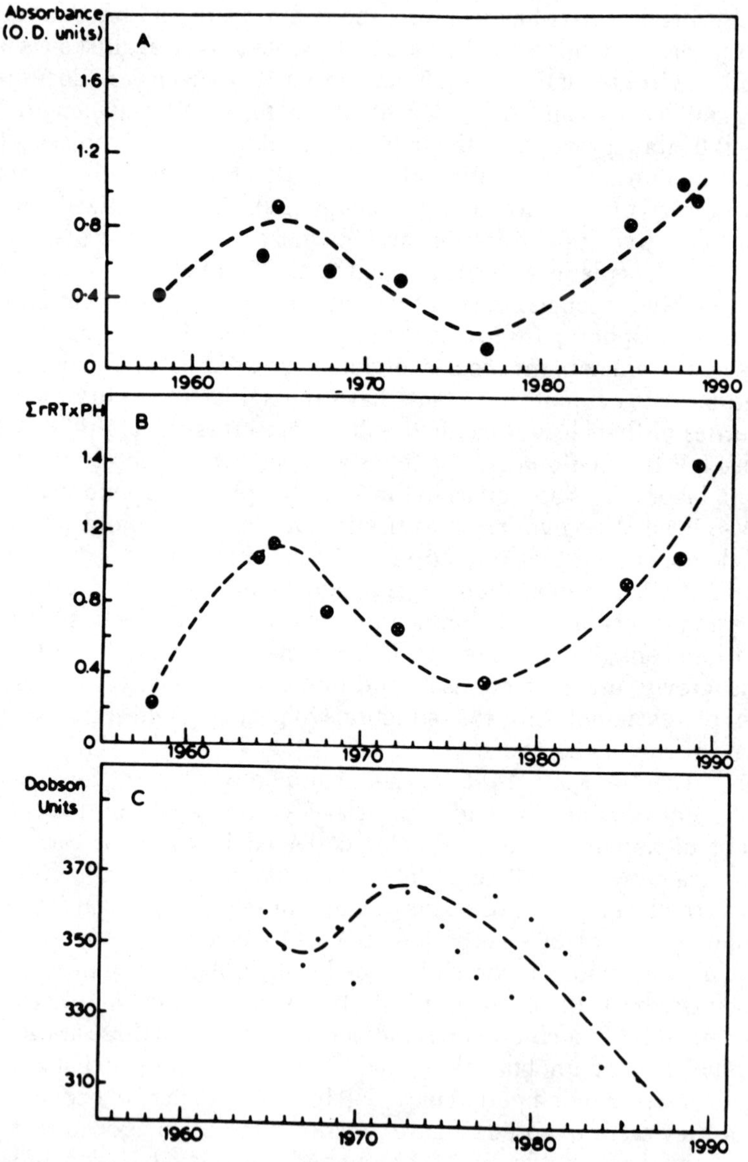

Figure 24. Ozone level trends in the Antarctic as indicated by flavinoid levels in moss. (A) Relative historical flavinoid levels in argenteum samples as determined by absorbance measurements. (B) Relative historical flavinoid levels in argenteum samples as determined by chromatographic methods. (C) Mean December ozone levels at the South Pole. (Courtesy of K. R. Markham, Chemistry Division, Department of Scientific and Industrial Research, New Zealand.)

that agents other than CFCs may be at work. Clearly, collections of moss specimens obtained prior to 1956 would indicate if ozone reductions occurred with any regularity. If ozone reductions were shown to occur from time to time it would suggest that possibly the CFC theory is overly simplistic. More importantly, the work on flavinoids suggests that it is premature to assume that a decline in the ozone layer automatically means destruction of living things receiving increased exposure. This idea receives further support from other investigations.

To gain greater insight into the possible effects of increased UV penetration on marine organisms, in particular phytoplankton, which are unable to synthesize flavinoids, the National Science Foundation sent a research vessel to Palmer Station in the Ross Sea to obtain the needed information. Deneb Karentz of the Laboratory of Radiobiology and Environmental Health, University of California, San Francisco, was one of the expedition's members. she has found that of the 57 organisms she collected, 86% possess a built-in "sunscreen" consisting of amino acids, which absorb potentially harmful radiation. These compounds are especially evident in those organisms inhabiting the upper reaches of the sea, where light penetration is greatest and potentially most harmful. In the deeper areas, there is a sharp decline in the presence of these amino acids, which may be related to the fact that radiation protection comes from the water itself. Penetration also falls off sharply with depth.

As for krill, the small shrimplike marine crustaceans that constitute the principal food of cetacean whales (right whale, blue whale, baleen), there is yet no objective evidence of harm. While it may be too early for final judgments, one would have expected some evidence of loss given the large holes in the ozone layer over the past several years.

Apparently, damage to DNA is not mortal. All these organisms can repair damaged DNA. At least three cellular mechanisms react with damaged DNA to restore the original genetic code: photoreactivation, excision repair, and recombinational repair. In the process of photoreactivation, phytoplankton having objective evidence of dimer formation as a consequence of UV radiation were able to rid themselves of the damage in a relatively brief period. Karentz found that light in the visible range enhanced DNA repair. It is worth recalling that these marine organisms have evolved over thousands of years in these waters and more than likely have developed appropriate defense mechanisms. Clearly, without resort to the organisms themselves, predictions of biological crises on the basis of observations of holes in the ozone layer and measurements of surface radiation are unwarranted. Nevertheless, the evidence that CFCs are related to reductions of the ozone layer is clear and direct. The fact that a growing body of information indicates that neither human health nor the Antarctic food chain has been adversely affected by that reduction is not sufficient reason to drop the fight for a ban on CFC production and use—until such time as their chemical structure can be modified. It would also be in our best interests to determine if natural events in the stratosphere also contribute to ozone depletion.

THE WATER BUDGET

Climate models employing enhanced greenhouse gas concentrations have projected increased winter precipitation of as much as 15% and decreases in summer rainfall of 5–10% in the central United States by 2030. Accordingly, summers could become hotter and drier with untoward consequences for cattle and crops. The almost singular concern for increasing global temperature often overlooks the potentially dislocating effects to the hydrologic cycle, the earth's water budget, and our water supply. Effects of excessively increased or decreased rainfall and corresponding increases or decreases in soil moisture could be far more damaging than temperature increases. Shifting patterns of precipitation could have a far greater impact on daily life, given the needs for drinking, flushing waste, cooking, cleaning, recreation, flowers, plants, crops, and livestock. Increases in global temperature imply greater evaporation from both the seas and land, which also means increased moisture in the air, simply because the capacity of air for water vapor increases some 5–6% for each 1°C increase. And added moisture can have two deleterious effects: some regions are predicted to become drier and some wetter, and some drier and hotter. However, the amount of rainfall can be a misleading indicator, because it is soil moisture rather than total precipitation that actually determines crop growth and subsequent agricultural activity. Concern for each of these was underscored when Manabe and Wetherald of the NOAA/Princeton University GFDL published the results of their model runs, which indicated that soil moisture "would be reduced in summer over extensive regions . . . such as the North American Great Plains, western Europe, northern Canada and Siberia" (25). Their model predicted that the soils in these areas could lose as much as half of their moisture content. Such large losses, they believed, would result from greater evaporation and this would be augmented by earlier melting of snow and thawing of soils, which would further enhance the rate of evaporation. Although not stated expressly, a loss of 50% of soil moisture implies that the Midwest could become a disaster area, a veritable Dust Bowl. Not everyone agrees. Among the questions raised is whether models account for the mechanism of retention of moisture in soil, and if they do not, can they be sure how a warmer atmosphere will affect binding of moisture to soil particles.

Figure 25 shows the responses of five major models to a doubling of CO_2 on soil moisture, in winter (Figure 25a) and summer (Figure 25b). Each model is seen to produce quantitatively different results. Nevertheless, all

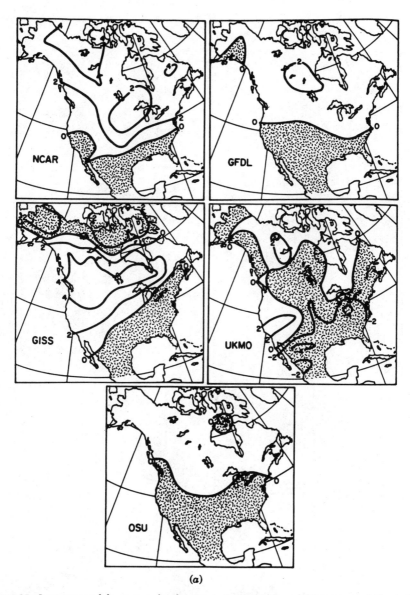

(a)

Figure 25. Increases and decreases of soil moisture in (a) winter and (b) summer relative to the control when CO_2 is doubled. The shaded areas are those where there is a decrease in soil moisture or change to a drier condition; clear areas show an increase. Units are centimeters of water.

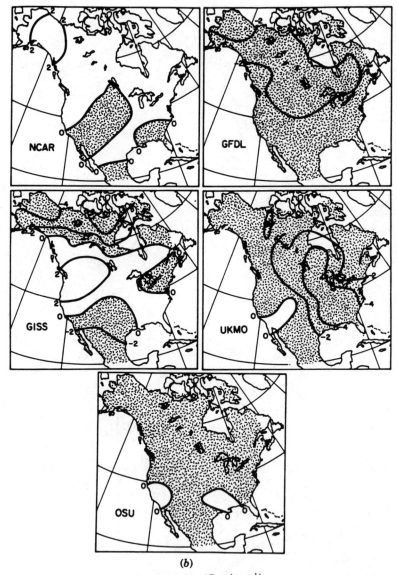

(b)

Figure 25. (Continued)

predict substantial warming in the summer months in similar areas of the continent, which would suggest they be given some credence. To their credit, however, Manabe and Wetherald raise a caveat and admonish readers that "a recent study by Somerville and Remer suggests that the increase in the liquid water content of clouds in response to the warming of air may act as a negative feedback between temperature and cloud cover by increasing the planetary albedo [reflectivity], thereby reducing the sensitivity of climate." This effect, they continue, "is not considered in this model. In view of the primitive state of the art for the parametrization of cloud formation, and other processes, the quantitative aspect of the present study should be interpreted with caution." The Manabe–Wetherald study is cited here because its predictions are the most dire and because Manabe and Wetherald are among the world's foremost modelers. Futhermore, Kellogg and Zhao of NASA and Peking University, respectively, who originally compared these five state-of-the-art climate models, had this to say: "The crucial question still remains, however, as to how well they actually simulate the 'real climate system.' Of more importance . . . we must also ask how well they simulate *the response to a change of boundary conditions or forcing factors* [italics theirs] of the climate system. . . . The first question can be partly answered by noting how well they simulate the present climate when all the boundary conditions are the same as now. It has been pointed out that in some respects they do very well . . . however, when it comes to calculations of precipitation, they do not seem to do so well. . . . Precipitation and soil moisture are secondary features of the general circulation, and depend on a complex sequence of interactions of flow patterns, storm tracks, convective activity, and so forth. It is not surprising, therefore, that the hydrologic cycle is more difficult to model than pressure and temperature distributions. . . . The fact remains that, when five competent and more or less independent research groups develop models that behave in approximately the same way, there should be occasion for some rejoicing. In the realm of precipitation and soil moisture, the time for unbridled rejoicing has not yet arrived. . . . Curiously, two or three of the models tended to agree very well in winter. . . . One must conclude that their high degree of confidence in their individual results, established statistically, was not justified in any absolute sense. Finally, it is our conviction that, in spite of the disappointing lack of original agreement between the various climate model calculations of changes in soil moisture on a warmer earth, there is some useful information in these results. One may conclude . . . in a preliminary way that the evidence points to some increase in soil moisture in winter . . . and some drier conditions in the southern states and Mexico. In the summertime, we can say, but with somewhat less confidence, that there may be some tendency for dryness in the middle of the continent, and the opposite may be true along the Gulf Coast and the West Coast of the United States" (26). At this point, it may be well to recall that great difficulties in forecasting can be expected because, in addition to temperature, greenhouse warming will, in all likelihood, alter cloud cover, wind velocity, and humidity, all of which play a role in determining the amount of moisture that air can accommodate. In addition, these changes will certainly affect plant growth and their ground covering characteristics, which in turn must affect evapotranspiration. If either precipitation or evapotranspiration or both change, the amount of water stored in the soil, the amount that penetrates beyond plant roots, and the amount that drains off to rivers and streams will also change. Given this panoply of unknown conditions, it must be seen as remarkable that these models are so strikingly similar.

Although, as Kellog and Zhao suggest, climate models are not yet equipped to forecast accurately, it does appear that, as a group, modelers tend to view a warmer world as afflictive for humankind. It is unfortunate that modelers have yet to develop their scenarios in conjunction with agricultural experts—horticulturists, soil scientists, hydrologists, and agricultural economists. This could be more sanguine. Seeing the world through rose-colored glasses, however, is not a prerequisite for forecasting, but the recent estimate of climate change and its impact on U.S. agriculture developed by a clutch of climatologist/modelers/agricultural scientists does set off the differences in sharp relief.

Richard M. Adams of the Department of Agricultural and Resource Economics, Oregon State University, and nine colleagues from universities and institutes around the country concluded their recent in-depth analysis somewhat tongue-in-cheek, stating that "the impact on the U.S. economy strongly depends on which climate model is used." Although one value does not a program make, it does serve to suggest the type of problem involved. Thus Adams et al. point out that, historically, monthly precipitation for Columbia, South Carolina averaged 113 mm during the growing season, but that, with scenarios employing doubled CO_2, the GISS model forecast 148 mm and the GFDL model predicted 50 mm. "The lower summer growing season precipitation for the GFDL model accounts for its much lower yield under rainfall conditions." They point out too that "the beneficial effect of elevated CO_2 on crop yields . . . offsets some or all of the adverse climate effects." And they make a salient point, suggesting that models should reflect "the adjustment that producers, consumers and other affected parties are likely to take to soften the impacts of adverse climate. Thus, human behavioral responses to adverse climate must be modelled." The idea that as we approach the millennium and enter the 21st century human beings would be seen as passive recipients of climate change does not speak well of modelers' scenarios. Adams and co-workers are, of course, correct in assuming that people everywhere will respond in any number of imaginative ways to create suitable living conditions. They always have. It's simply a human trait. Numeric models deal with what they know best: numbers. Thus far, there has been no place for emotions and human responses. Their conclusion leaves no room for equivocation. "Major policy concern is whether climate change is a food security issue for the United States. The result of these analyses suggests that it is not" (27). They are also reasonably confident that loss of rainfall in certain areas could be made up by increased irrigation. Unfortunately, the question not raised concerns from where additional water resources would come.

This leads inexorably to the concern for water resources, the supply available for use. Figure 26, the hydrologic cycle in a global view, and

Figure 26. The hydrologic cycle: a global view. (Courtesy of the U.S. Geological Survey, U.S. Department of the Interior.)

Figure 27, a more manageable view, describe the constancy and balance of the earth's current total water budget. The same water has been transferred, time and again, from the oceans to the atmosphere, dropped upon the land, and transferred back to the sea. This endless cycle is known as the hydrologic cycle. It is a natural machine, a constantly running distillation and pumping system for which the sun supplies heat energy and, together with the force of gravity, keeps the water moving. As shown in the figures, the movement is from the earth to the atmosphere as evaporation and transpiration, and from the atmosphere to the earth as condensation and precipitation, and between points on earth as stream flow and groundwater movement. As a cycle, the system has neither beginning nor end. From the human point of view, the oceans are the major reservoir or the source, the atmosphere is the supplier, and the land is the user. In such a system, no water is lost or gained, but the amount available for use may fluctuate because of variations at the source or, more often, in the atmosphere. Climate shifts in the geologic past have produced ice ages with a great drawing up of water into ice with consequent lowering of the seas, and it has made the land flower or dry up into deserts across entire continents. Currently, small alterations of the local patterns produce floods and drought.

Seventy-two percent of the earth is covered by water. That translates into 361,637,000 square kilometers (139,628,000 square miles), and 97% of

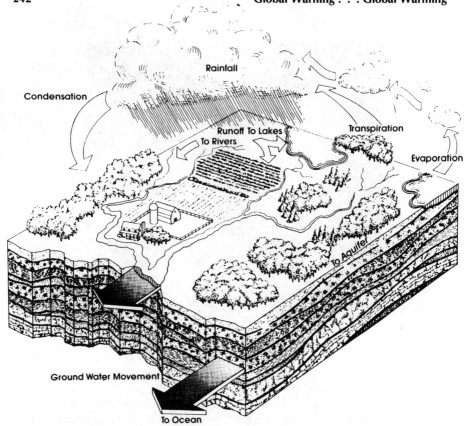

Figure 27. The hydrologic cycle: a local view.

all surface water is in the world's oceans. Some 3% is "fresh" or nonsalty. Much of the water remains unavailable, locked away as it is in the ice of glaciers and the great ice sheets of the polar regions. At any instant, about 0.005% of the total supply is in motion, moving through the water cycle. In the United States, a drop of water spends an average of 12 days passing through the air, or it may remain in a glacier for 40 years, in a lake for 100 years, or in the ground from 200 to 10,000 years. Essentially, every molecule moves through the hydrologic cycle. In 1 year, some 95,000 cubic miles* of water evaporate and approximately 75,000 cubic miles return. But is should be recalled that the total supply has not grown or diminished, nor should it be expected to, unless and until, in a warmer world, the distribution of solar energy shifts the balance. In geologic terms, from our Holocene to the Pliocene epochs, the cycle has continued unabated. Thus the water we drink today was contained in barrels carried on the Santa

*To obtain an idea of the tremendous gallonage involved, 1 cubic foot of water contains 7.5 gallons. To convert 95,000 cubic miles (mi³) to gallons, it is necessary to multiply by $(5280)^3$ and then by 7.5.

Maria, Pinta, and Nina, when Columbus and his crew sailed west to the new world. This endlessly recycled supply was the same conceived by Caesar and Cleopatra as they drifted leisurely down the Nile; and hunter-gatherers and artists of the upper Paleolithic period more than 30,000 years ago, who lived in and painted the cave walls at Altamira in Spain and Lascaux in France, drank the same water. We have all been drinking and using the same water, and, in this sense, we are all connected and continuous.

What can be expected of this water resource in a warmer world? Will a modified atmosphere affect the supply, and by what degree? In short, will water supplies be reliable?

To what can future projections be compared Figure 28 shows the total current water budget of the 48 continental United States as estimated by the U.S. Geological Survey (USGS). All values are expressed in billions of gallons per day (bgd). One bgd is an amount of water that would cover 1 square mile to a depth of approximately 5 feet—4.8 feet to be more precise. If the U.S. Geological Survey is correct, it appears that 4200 bgd fall as rain and snow, which translates to 30 inches of average annual precipitation. But this is not evely distributed, and before it can even be thought of as available for use, evapotranspiration—the simultaneous loss of moisture from the soil surface along with the transpiration of water for green plants—returns 2800 bgd or 19 of the 30 inches. According to the USGS estimates, 1300 bgd run off to the seas, leaving 100 bgd for all human and animal uses.

Although 100 bgd seems to be a great deal of water, concern for an available supply—and available is the operative concept—is complicated

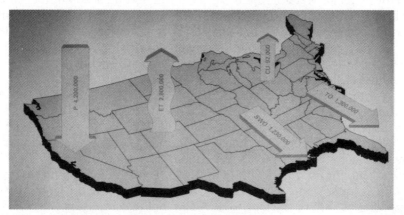

Figure 28. Estimated water budget, in millions of gallons per day, for the continental United States. Abbreviations: CU, consumptive use; ET, evapotranspiration; P, precipitation; SWO, surface-water outflow to the oceans, Canada and Mexico; TO, total surface-water and ground-water outflow to the oceans, Canada and Mexico. (Courtesy of the U.S. Geological Survey, U.S. Department of the Interior.)

not only by varying population density but by differences in point of supply and point of use. Such potential asymmetry is nowhere more realized than in California, where 75% of its annual precipitation falls as snow in the Sierra Mountains to the north, while 80% of the state's total supply is consumed in the spring and summer in its central valley and southern countries.

In the southwestern and western states, water shortages have been chronic and a problem for at least several decades, and they are expected to worsen as population shifts to the sun belt states. In a warmer world, evaporation, as noted earlier, must increase. This implies an increase in the 2800-bgd evaporative level to a higher but unknown and uncertain level, which would further reduce the 100-bgd human/animal/plant portion. Unless we know future precipitation levels, water resources will remain speculative and uncertain.

If, as discussed earlier, crops become more photosynthetically efficient, water needs and use would increase with the net result being no different. However, the uncertainties are so great that predictions become almost meaningless.

It is this uncertainty that hampers decisionmaking on such formidable questions as whether large reservoirs or catchment basins should be built in areas where heavy rainfall is predicted. Reservoirs with pipelines capable of carrying water hundreds or thousand of miles to water-short areas could alleviate potential dislocations, but construction costs would easily run to tens of billions of dollars. On what basis can such a decision be made? Given the prevailing uncertainties, would such a decision by justified? A case in point, again from California may serve to illustrate the complexity of the issue.

With its major water supplies in the north and its needs in the south, California built itself one of the world's largest reservoirs to store and distribute water as needed, but especially to its central valley, one of the world's most productive agricultural regions, which produces more than 50% of all fruits and vegetables consumed in the United States.

Located at the foot of the Sierra Nevada Mountains, 75 miles north of Sacramento, the Oroville/Thermalito complex is a core water storage and electrical generation facility for the California State Water Project. Lake Oroville covers 14,810 acres and hold 3,538,000 acre-feet of water—enough to supply more than half of California's urban water needs for a year. One acre-foot is equivalent to about 326,000 gallons of water, the average amount used by a family of five in 1 year. Impounding this vast amount of water is the 770-foot high Oroville dam, one of the largest in the United States. Although the Oroville/Thermalito complex is a major unit of the State Water Project, the Water Project itself consists of over 150 dams and reservoirs, built to hold water as a hedge against drought: unless, of course, drought persists and there is little water to be held, which is the current state of affairs.

In a warmer world, however, snow in the Sierras would melt earlier, adding runoff to an already dicey winter water management problem. To protect their capital city of Sacramento from flooding, reservoir levels would have to be drawn down in the winter. Such water reductions would mean less water available for the south and central valley in summer. The usually adequate summer irrigation requirements would become highly uncertain, and with higher summer temperatures, the demand for drinking and recreational water, along with greater crop needs, would exacerbate the shortage even more. An increased number of warmer days would call for increased use of air conditioning and, with it, increased demand for electric power, which would require additional supplies of cooling water for the increased number of generating plants. Faced with this uncharted cycle of demand, do they build additional reservoirs around the state or do they look elsewhere to augment their water resources? For example, would California and other hard-pressed western states look to the Great Lakes for additional water? Should the five Great Lakes be considered as a safety valve for threatened areas? Would it be more important to divert water from the Great Lakes to the Mississippi River to assure uninterrupted passage of river traffic? Would Lake states, with state sovereignty, countenance such tapping and shifting? More than likely, such attempts would be stalled in the courts for years. Perhaps other solutions should be sought. What about conservation? With increased temperature and a water supply shortfall, watering lawns, swimming pools—which California has more of per square mile than any other area of the world—cleaning streets, car washing—which, in California, accounts for immense water use—would become casualties. Should "gray" water—that is, waste water—be reprocessed, "polished," cleaned, and reused? There is adequate precedent and experience for this, but a public raised on cheap water and with wasteful habits may balk. Nevertheless, it is preferable to going without, and the amounts saved can easily run to millions of gallons per day. Another case in point is the current water shortage in Israel, where waste water is reused. Some 180 million cubic meters of sewage are currently being recycled. The flow from seven towns in the Dan region is collected and polished by biological waste treatment, then used in part to recharge aquifers and irrigate commercial crops such as cotton. There is a pertinent message here for California. Conservation, with or without reuse, could become a stabilizing factor. There is no lack of alternatives. But decisions will have to be made before "the weather turns."

California is faced with twin problems simultaneously: the worst drought in 40 years and relentless population growth. Eighty-five percent of all available water has regularly been used to irrigate crops, including such water-demanding ones as alfalfa, cotton, and rice. Exacerbating the problem is the often profligate use of cheap water, which has led to great waste. The pricing policies of the Bureau of Reclamation have encouraged both the wasteful use of water and the unwarranted expansion of irrigated land—producing a classic example of the Tragedy of the Commons, wherein it can be considered in each farmer's best interest to place just a few more acres under irrigation, but also wherein increasing the total acreage under irrigation is also to everyone's ultimate detriment. The remaining 15% is used for drinking and other needs. But the ever-westward trek of a shifting population, along with new immigrants over the past 10 years, has increased its population by a hefty 26%; from 23 million in 1980 to 30 million in 1990. This has also played hob with the supply. The unbalanced use of the supply has now led to recrimination and public outcry by urban dwellers, who believe they have been ill used by farmers who have grown rich while wasting precious water.

Together, cotton and alfalfa use 25% of the available 85%. Alfalfa alone requires more water per acre than any major crop, except rice, and thus uses more water than the combined household needs of California's 30 million residents. In 1990, the alfalfa crop used fully 16% of the total water supply. It does give pause.

To further aggravate the situation, over 15 million people have shoe-horned themselves into the area from Ventura, 30 miles north of Los Angeles, to the Mexican border, all depending on water from the far north. Population and town growth have been wholly uncontrolled and little to no planning for such eventualities has taken place. It is the lack of plan-ning, this waiting until a crisis occurs, that is a national characteristic, which bodes ill for reducing the effects of global warming in time.

Although model predictions and projections can suggest where and when precipitation patterns may be adversely or beneficially affected, it is well beyond their scope to suggest how the available supply should be used. People will determine that. Models can be a source of information, which people choose to use or not use. And California's use of its available water may become a cautionary tale for the midwestern and southwestern states.

It is of further interest to recall that model projections suggest, as noted at the outset, an increase in winter precipitation and a decrease in summer precipitation. But it is also well to recall that projections of a modified pattern of precipitation require that the past patterns be known. Given the lack of such information across the oceans of the world, which represent 72% of the earth's surface, that pattern will remain largely unknown. Fortunately, however, on certain regional levels, that information can be addressed. For example, tree-ring chronology can assist in deriving past patterns. Is the protracted drought in California a consequence of global warming? Perhaps. But drought, an all too common occurrence, is a natural phenomenon. Recently, Blasing, Stahle, and Duvick using tree-ring chro-nologic assessment performed a 231-year reconstruction of annual precipi-tation from 1750 through 1980. They found that drought appears to occur at intervals of from 15 to 25 years, and that the 1950s drought was the most severe of the 20th century, but that the drought of 1860 in the south-central

United States was probably the worst in the last 231 years. Possibly of most interest is their contention "that drought of greater severity and duration than any recorded can occur whether or not a CO_2-induced warming, or any warming, occurs" (28). Obviously then, great caution is called for in asserting that a lack of precipitation or a period of drought can be ascribed to a CO_2-induced warming. It is essential that knowledge of the cause dictate policy.

Traditionally, of course, fresh water has been the water of agriculture the world over. But in many countries, both prime farmland and fresh water are fully utilized. How then can food production expand to feed a growing population? In a recently published survey, the National Research Council found that, around the world, literally hundreds of plants can be successfully grown in seawater (28). Salt-tolerant plants (halophytes) can use land and water otherwise unsuitable for conventional crops (glycophytes) for the extensive production of food, fodder, and fuel. Halophytes growing in soils or water containing substantial amounts of inorganic salts can use saline environments that are currently neglected and widely considered impediments, rather than opportunities for development.

Salts occur naturally in all soils. Rain dissolves these salts, which are then swept through streams and rivers to the sea. Where rainfall is sparse, some of this water evaporates and the dissolved salts become more concentrated, which can result in the formation of salt lakes or in brackish groundwater, salinized soil, or salt deposits.

In some developing regions, there are millions of acres of salinized farmland resulting from poor irrigation practices. These lands would require large amounts of water—water that does not exist—to leach away the salts before conventional crops could be grown. However, there are salt-tolerant plants that can be grown on them without expending precious fresh water to wash away the salts.

From experience in irrigated agriculture, Miyamoto suggests the following classification of potential crop damage from increasing salt levels:

Irrigation Water	Salts (ppm)	Crop Problems
Fresh	<125	None
Slightly saline	125–250	Rare
Moderately saline	250–500	Occasional
Saline	500–2500	Common
Highly saline	2500–5000	Severe

Figure 29 shows the type of growth response to increasing concentrations of salt. Evidently, the growth of halophytes is improved by available salt. Other crops, such as barley, grow well at low salt concentrations but beyond an optimum level growth is precipitously reduced. With salt-

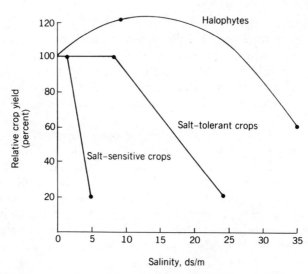

Figure 29. Growth responses to varying degrees of salinity. Some halophytes have increased yields at low salinity levels. Salt-tolerant crops maintain yields at low salinity levels but decrease yields as salt levels exceed a certain limit. Salt-sensitive crops decrease yield even in the presence of low salt levels.

sensitive plants, such as beans, yields decreased rapidly with increasing increments of salt.

There is little doubt that salt-tolerant species could readily "improve food or fuel supplies, increase employment, help stem decertification and contribute to soil reclamation." Perhaps even more important is what the report doesn't say. In a warmer world, specifically in those locations predicted to be drier, and where salt buildup can be expected to occur, halophytic plants could be exploited successfully. They are not limited to developing countries. It is of interest to learn that most of the 1500 plants covered in the NAS report had been used in the past. These are not merely developed species. The Seri Indians of the American Southwest once pounded the seeds of the seagrass *Zostera marina* into flour, while Palmer's saltgrass was harvested from the tidal flats of the Gulf of California to make flour for bread. Today, in Israel, saline water is used to irrigate commercial crops of cotton and tomatoes, and in Mexico farmers currently farm Salicornia, a succulent plant for its safflower-like oil. Perhaps even more far-reaching in it potential benefits is the distinct possibility of transferring genes for salt tolerance and drought resistance from halophytes to such glycophytes as aspargus, wheat, barley, and rice, further reducing the untoward effects of a predicted warming.

EFFECTS OF CLIMATE CHANGES ON HUMAN HEALTH

By enriching the atmosphere with CO_2 and other IR absorbing gases, we human beings may have inadvertently added a unique source of risk to our lives. It is not the gases that may be the problem, but the temperature increase, which brings with it the stress of added heat. If computer models are accurate, warming will be more evident in the winter than in the summer, particularly, but not limited to, the higher latitudes, 50–60°N latitude and above. Figure 30 is a model-derived representation of the spatial patterns of surface temperature differences between cold and warm-year country groupings, as well as a representation of differences in annual mean temperature. Maximum warming is seen to occur in the high latitudes, above 65°N, by as much as five times more than the hemispheric mean. This area also coincides with the region of greatest natural temperature variability.

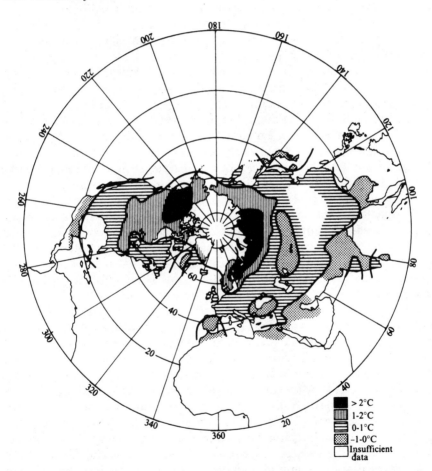

Figure 30. Mean annual surface temperature from cold to warm years. The corresponding change in the hemispheric mean temperature is 0.6°C. For reference, the expected change in global mean temperature due to a doubling of atmosphere CO_2 concentration is approximately 2°C. Most of the large temperature differences are seen to occur at the higher latitudes and also appear to occur most often in winter. Data are based on temperatures obtained at 219 land-based stations. Reprinted by permission from *Nature*, Vol. 23, pp. 17–21, 3, Copyright (c) 1980 Macmillan Magazines Ltd.)

But that doesn't mean that summers will not be hotter; they will, in a warmer world. Winter will not merge with summer. We shall continue to have our seasons. And warmer winters could be an unpremeditated bonus for many who dote on such winter sports as skiing, bobsledding, skiboarding, and other risky, usually high-speed downhill activities, in that fewer broken legs, arms, and necks would be collected between the winter solstice and the vernal equinox.

Temperature increases may affect us directly if there is prolonged exposure with failure of heat loss mechanism, or it may alter the relationship between man and microbes; that is, human resistance may decrease, the virulence of microorganisms may increase, or a changing environment may bring man and infectious agents closer together. Although human illness and the environment have traditionally been seen as inextricably linked, we may have shed that relationship with the passage of an agrarian way of life. The idea that environmental factors can affect our health is as old as . . . as mankind. The traditions of primitive peoples across the globe reflect the ill effects of certain winds, then changing of the seasons, the phases of the moon, and the influence of the sun and stars.

A cogent delineation of the unit of our habitat written some 2500 years ago, but whose message may no longer be as vital, was set forth by Hippocrates in his treatise *On Airs, Waters, and Places*. As this 4th century B.C. physician saw it, "whoever wishes to investigate medicine properly should proceed thus: in the first place to consider the seasons of the year and what effects each of them produces. Then the winds, the hot and the cold, especially such as are common to all countries, and then such as are peculiar to each locality. In the same manner, when one comes into a city to which he is a stranger, he should consider the situation, how it lies as to the winds and the rising of the sun; for its influence is not the same whether it lies to the north or the south, to the rising or to the setting sun. One should consider most attentively the waters which the inhabitants use, whether they be marshy and soft, or hard and running from elevated and rocky situations, and then if saltish and unfit for cooking; and the ground whether it be naked and deficient in water, or wooded and well watered, and whether it lies in a hollow, confined situation, or is elevated and cold; and the mode in which the inhabitants live and what are their pursuits, whether they are fond of drinking and eating to excess, and given to indolence, or are fond of exercise and labor, and not given to excess in eating and drinking. . . . If one knows all these things, or at least the greater part of them, he cannot miss knowing when he comes into a strange city, either the diseases peculiar to the place, or the particular nature of common diseases, or commit mistakes, as is likely to be the case provided one had not previously considered these matters."

Clearly, we are not totally detached from our environment. Nevertheless, as Table 4.5 shows, the leading causes of death in the United States and western Europe appear not to foster such a relationship. This does not

Table 4.5
Leading Causes of Death in the U.S., 1988 (per 100,000 Population)

Rank	Number of Deaths[a]	Percent
1. Diseases of the heart	317.5	36.0
2. Malignant neoplasms	194.7	22.0
3. Cerebrovascular disease (stroke)	62.1	7.0
4. Accidents	39.5	4.4
5. Chronic obstructive lung disease	31.8	3.6
6. Pneumonia/influenza	29.0	3.2
7. Diabetes	15.4	1.7
8. Suicide	12.8	1.4
9. Liver disease/cirrhosis	10.9	1.2
10. Atherosclerosis	9.4	1.0
11. Homicide	9.0	1.0
All others	152.1	17.2
Total (all causes)	884.0	100%

[a] Deaths per 100,000 population.

mean that environmental factors such as heat, light, wind, pressure, and moisture no longer affect us. But the benefits of technology, air conditioning, refrigeration, and heating systems have rendered these environmental factors decidedly less intrusive. Whereas excessive heat can overtax the heart and coronary vessels, it is not usually a primary cause of heart disease deaths. The very old, the very young, and the chronically ill may succumb in hot weather, but excessive heat will more than likely be a secondary rather than an underlying cause of death. Be that as it may, the most trying environmental stress for most people will be increasing summer heat, especially if accompanied by high humidity. Among the predictions of a warmer world are increased number of days with daily summer temperatures above 85 and 90°F along the Atlantic seaboard, from Boston to Washington, DC and inland in cities such as St. Louis and Dallas, where high humidity has always been a fact of life in summer. Increased numbers of high heat days will undoubtedly lead to excess deaths among infants, the elderly, and the chronically ill, if adequate air conditioning is unavailable. This is not conjecture or speculation, nor are a model and a computer necessary to obtain potential numbers of expected excess deaths. The data are already at hand. Heat-related deaths have occurred every June, July, and August, especially when the heat hangs on. As we shall see, it's the hanging on that does it.

It is noteworthy that, since the 1960s, both the incidence of and death rates from heart disease have fallen substantially, while temperatures have been rising—according to model estimates—especially during the decade of the 1980s, the warmest decade of the century. Few of the other conditions listed in Table 4.5 lend themselves to heat as a primary cause of death. In the heat and humidity of summer, particularly in the inner cities where air conditioning is often lacking, irritation over a real or imagined slight or injury could lead to one person killing another. Few, if any, have taken their own lives because of oppressive heat. But as noted earlier, it is the prolonged exposure to high heat often accompanied by high humidity that places the body in jeopardy. Prolonged exposure may lead to either excessive fluid loss and heat exhaustion—a failure of the heat loss mechanisms—with consequent heat stroke. Dehydration, age, excessive sweating, vomiting, diarrhea, and debility predispose to either; high humidity, strenuous exertion, poor ventilation, and heavy clothing also contribute to the possibility of developing dehydration or heat exhaustion.

Heat stroke can be the most serious health problem in a hot environment. It occurs when the human thermoregulatory system ceases to function and sweating stops. Why this occurs remains a physiological conundrum, and most distressing is the fact that the individual involved is unaware that it is happening. But when it does, the body's only effective means of ridding itself of excess heat is gone. The skin becomes hot and dry to the touch and reddens dramatically. At this point, body temperature is at least 104°F and rising. Mental confusion is another early sign. If, at this point, a person is not quickly cooled, the end stages are unconsciousness, delirium, convulsions, and death.

A lesser degree of heat stress is heat exhaustion. This proceeds from excessive loss of fluids and/or salt. Sweating continues to occur, but there is muscular weakness, fatigue, often giddiness, and headache. The key to heat exhaustion is a moist, clammy skin, but body temperature usually remains normal. Cooling and fluids are called for to bring relief.

Heat cramps, at the far end of the heat stress spectrum, are the pain of muscles in spasm. This is most often seen in those who sweat a great deal, consume copious amounts of fluid, but fail to replace lost salt. Imbibing of fluid without salt further dilutes the body's salt (electrolyte) balance. It is the loss of salt from muscle which induces the painful cramping. People on low- or no-salt diets should be made aware of the forms of salt other than sodium-based which are suitable for their special needs.

Human beings are adaptable and capable of adjusting to high heat. Acclimatization takes a week or two, but the body does undergo changes that make exposure to higher heat levels more endurable. In a warmer world, the operative term is common sense, which, as my grandmother attempted to make me aware, is most uncommon. Hence heat-related deaths can be expected to remain an integral part of the summer season.

Then there are the more indirect effects of a temperature increase. Here we must consider the possible change in the relationships between humans and microbes. Special attention should be given to the vecter-borne diseases, that is, those conditions listed in Table 4.6, in which arthropods, such as ticks, and mites, and insects, such as mosquitos, are involved. The geographic distributions of tick-borne diseases—the spotted fevers and Lyme disease—are affected by climatic factors. Such variables as tempera-

Table 4.6
Selected Diseases in Which Arthropods and Insects Are Linked

Diseases Linked to Arthropods	Diseases Linked to Insects
Colorado tick fever	Malaria
Lyme disease	Western equine encephalitis
	Dengue fever
Scabies	Trypanosomiasis
	Rift Valley fever
Typhus fever	Eastern equine encephalitis
	St. Louis encephalitis
Rocky Mountain spotted fever	Semliki-Forest disease

ture, humidity, and vegetation directly affect the sizes of both the tick population and the deer, mice, and bird populations, the means (vectors) of bringing ticks into contact with people. Mosquitos are sensitive to weather, often preferring warm, humid environments. The spread of mosquito populations and the disease agents they carry—protozoans (malaria) and viruses (yellow fever and various encephalitides)—depends in part on temperature, humidity, and available vegetation, all influenced by climate.

Mosquito-borne diseases have been well controlled and thus, for the most part, prevented in the United States over the past 20 years. State-wide programs for mosquito control provide constant surveillance of mosquito populations and should continue to function well. If populations of ticks should increase as a result of climate shifts, available, in-place control measures can reasonably interdict at least one of the several links in the chain of causation. Figure 31 illustrates the multiple links involved in a typical arthropod-borne disease. Although more complex than other diseases, they also offer a number of points at which control can occur.

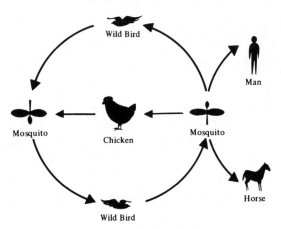

Figure 31. Arthropod-borne diseases usually have a number of links in the chain of transmission.

Chapter 3 and this chapter have attempted to delineate the current state of knowledge and understanding of the many-faceted issue designated "global warming." The facets are inordinately complex and fraught with inherent uncertainty because neither prior experience nor guidelines of any type exist on which investigators of any discipline can base their ideas or compare data. The admonition that there is nothing new under the sun no longer applies; there is something new for which there is no previous human experience and the only reasonable way to study the problem is with the use of computer-generated mathematical models, which are themselves limited in scope because the knowledge base of the subject being simulated is lacking. In the face of this void, scientists in a range of disciplines are attempting to fill in the background by a combination of laboratory studies, limited field trials, and intuition. Which is fine as far as it goes. But there is an uncertain and unsatisfactory state of affairs which is no where more forthrightly purveyed than in the recent Policy Makers Summary of the Scientific Assessment of Climate Change, the report of the Intergovernmental Panel on Climate Change (IPCC) (30). This should be widely read—but slowly and with an eye for the nuance of language.

What do we know for certain? they ask; to which the response is:

- We know that without a natural greenhouse effect in the atmosphere the earth would be colder than it is.
- We know too that emissions from the combustion of fossil fuels have substantially increased in the concentration of greenhouse gases, which will enhance the greenhouse effect, "resulting on average in an additional warming of the earth's surface."
- We know that in response to the warming, the atmosphere will be enriched in water vapor, which will further enhance warming.
- We know that chlorofluorocarbons are persistent chemicals.

It was also the judgment of the panel that, with respect to the reported 0.3–0.6°C increase in temperature over the past 100 years, "the observed increase could be largely due to natural variability; alternatively this variability and other human factors could have offset a still larger human-induced greenhouse warming. The unequivocal detection of the enhanced greenhouse effect is not likely for a decade or more."

That's what is known with certainty. For regional changes, there was a lesser degree of confidence. For example, estimates of changes predicted to occur in five regions of the world were as follows:

ESTIMATES FOR CHANGES IN CLIMATE BY 2030

The numbers presented in the following discussion are based on high-resolution models, scaled to be consistent with our best estimate of a mean global warming of 1.8°C by 2030. For values consistent with other estimates of global temperature rise, the numbers below should be reduced by 30% for the low estimate or increased by 50% for the high estimate. Precipitation estimates are also scaled in a similar way. Confidence in these regional estimates is low.

1. *Central North America (35–50°N, 85–105°W)*. The warming varies from 2 to 4°C in winter and 2 to 3°C in summer. Precipitation increases range from 0 to 15% in winter, whereas there are decreases of 5–10% in summer. Soil moisture decreases in summer by 15–20%.

2. *Southern Asia (5–30°N, 70–105°E)*. The warming varies from 1 to 2°C throughout the year. Precipitation changes little in winter and generally increases throughout the region by 5–15% in summer. Summer soil moisture increases by 5–10%.

3. *Sahel (10–20°N, 20°W–40°E)*. The warming ranges from 1 to 3°C. Area mean precipitation increases and area mean soil moisture decreases marginally in summer. However, throughout the region, there are areas of both increases and decreases in both parameters.

4. *Southern Europe (35–50°N, 10°W–45°E)*. The warming is about 2°C in winter and varies from 2 to 3°C in summer. There is some indication of increased precipitation in winter, but summer precipitation decreases by 5–15%, and summer soil moisture by 15–25%.

5. *Australia (12–45°S, 110–115°E)*. The warming ranges from 1 to 2°C in summer and is about 2°C in winter. Summer precipitation increases by around 10%, but the models do not produce consistent estimates of the changes in soil moisture. The area averages hide large variations at the subcontinental level.

To his credit, Jerry D. Mahlman, a member of the Working Group responsible for preparing the IPCC Report and director of the GFDL, commented candidly that "the things we can say with confidence the policy makers are not interested in; and the things they are interested in, we don't know with confidence" (31). All else is speculative and commentary. Given such a milieu, it is not difficult to understand why President Bush and an administration committed to deregulation and a free market eschew any policy remotely smacking of curtailment of fossil fuel use. It is understandable but not entirely convincing. With the certainty that CO_2 emissions are rising, the Keeling CO_2 data are beyond reproach: warming of some level is entirely possible. That alone remonstrates against doing nothing. And there is a further consideration. Paleontologists have become unusually restive of late. For years they believed there would be no way to determine why our anthropoid ancestors left their treed preserves and took to walking upright. Now there are stirrings of a relationship between climate change and a change in species about 2.5 million years ago. "There seems to be a growing consensus" for a link between climate and human evolution. The idea that ecological disruption forced by a cooling trend brought apes out of the trees is no longer seen as farfetched. With a cooling trend and consequent loss of forest to plain and savannah, apes would need to forage further for food and doing so on two legs would be more efficient and faster and would free-up hands for grasping and holding. The evolutionary pressure to accomplish this would also develop a larger brain to handle the increased intelligence needed for the new tasks. This is an imaginative and appealing concept but that climate change was a cooling and it occurred over millions of years on an earth devoid of people, factories, skyscrapers, homes, and political boundaries. *Homo erectus* could stretch his legs as well as his imagination. Today, we are literally hemmed in on all sides and climate change, should it occur, may well be abrupt. Mankind will not be evolving, but changes would be forced on populations, and migration may be nigh impossible. These strictures suggest violent times if preparations are left for the moment.

It should be recalled that this shift in climate to a superinterglacial will, in geologic time, be extremely short; probably lasting no more than a thousand years—20 generations: enough time for the carbon cycle to absorb the excess carbon dioxide. After that, the earth's climate system should be back on track and moving into a cooling period as the interglacial declines into the next scheduled glacial period. Thus for humankind the climate problem is double-edged: preparation to thwart an unusual warm period, followed by preparations for a cooling period.

Time, always the enemy, is what is needed. Thus the recent model prediction of Michael E. Schlesinger and Xing Jian of the University of Illinois suggests that it is available (31). If over the next 10 years strict controls were deferred, any increase in potential warming would be minimal. Controls deferred for a decade would still achieve a 95% reduction in warming when they went into effect. Clearly, then it is in our best interest to avoid the type of abrupt curtailment of fossil fuel use which could produce worldwide economic destabilization, and at the same time introduce a broad-based program of energy efficiency that over the next 15 years could achieve a 15–35% reduction in emissions from 1987 levels, depending on the vigor with which the programs are pursued. Additionally, the decade of grace would give climatologists time to obtain the type of information needed to substantially reduce the uncertainty of their current predictions. Thus there seems to be sufficient time for everyone to benefit. Given the time, what can be done? Potential solutions are the message of Chapter 5. Therein are described those reasonable things that rational people can do, should do, to relieve a potentially dislocating problem. The keys are reasonable and rational.

REFERENCES

1. Franklin, B. Meteorological Imaginations and Conjectures. *Mem. Lit. Philos. Soc. Manchester* 2:373–377, 1785.

2. Sigurdsson, H., D'Hondt, S., Arthur, M. A., Bralower, T. J., Zachos, J. C., Fossen, M. V., and Channel, J. E. T. Glass from the Cretaceous/Tertiary Boundary in Haiti. *Nature* 344:482–487, 1991.

3. Bentley, C. R., and Giovinetto, M. B. Mass Balance of Antarctica and Sea Level Change. Personal communication, October 30, 1990.

4. Robinson, D. A., and Dewey, K. F. Recent Secular Variations in the Extent of Northern Hemisphere Snow Cover. *Geophys. Res. Lett.* 17:1557–1560, 1990.

5. Emanuel, K. A. Toward a General Theory of Hurricanes. *Am. Sci.* 76:371–379, 1988.

6. Emanual, K. A. The Dependence of Hurricane Intensity on Climate. *Nature* 326:483–485, 1987.

7. Gray, W. M. Strong Association Between West African Rainfall and U.S. Landfall of Intense Hurricanes. *Science* 249:1251–1256, 1990.

8. Davis, M. B. Insights from Paleoecology in Global Change. Address of the Past President. *Ecol. Soc. Am. Bull.* 70:222–228, 1989.

9. Gear, A. J., and Huntley, B. Rapid Changes in the Range Limits of Scots Pine 4,000 Years Ago. *Science* 251:544–547, 1991.

10. Wittwer, S. H. Carbon Dioxide and Climatic Change: An Agricultural Perspective. *J. Soil Water Conservation* 35:116–120, 1980.

11. Kimball, B. A. Adaptation of Vegetation and Management Practices to a Higher Carbon Dioxide World. In: B. R. Strain and J. D. Cure (Eds.). *Direct Effects of Increasing Carbon Dioxide on Vegetation*, Chap. 9. U.S. Department of Energy, DOE/ER-0238, Dec. 1985. NTIS, Springfield, VA, 1985.

12. Idso, S. B. Personal Communication. September 24th, 1990.

13. Wittwer, S. H. Carbon Dioxide Levels in the Biosphere: Effects on Plant Productivity. *CRC Crit. Rev. Plant Sci.* 2(3):171–198, 1985.

14. Fuller, J. G. *The Day of St. Anthony's Fire*. Macmillan, New York, 1968.

15. Matossian, M. K. *Poisons of the Past: Molds, Epidemics and History*. Yale University Press, New Haven, 1989.

16. Woodham-Smith, C. *The Great Hunger*. Harper & Row, New York, 1962.

17. Blumenthal, C. S., Barlow, S., and Wrigley, C. Global Warming and Wheat. *Nature* 347:235, 1990.

18. Blumenthal, C. S., Batey, I. L., Bekes, F., Wrigley, C. W., and Barlow, E. W. R., Gliadin Genes Contain Heat-Shock Elements: Possible Relevance to Heat-Induced Changes in Grain Quality. *J. Cereal Sci.* 11:185–187, 1990.

19. Srab, Z., Hajdukiewicz, P., and Maliga, P. Stable Transformation of Plastids in Higher Plants. *Proc. Natl. Acad. Sci. USA* 87:8526–8530, 1990.

20. Scotto, J., Cotton, G., Urbach, F., Berger, D., and Fears, T. Biologically Effective Ultraviolet Radiation: Surface Measurements in the United States, 1974 to 1985. *Science* 239:762–764, 1988.

21. Blumthaler, M., and Ambach, W. Indication of Increasing Solar Ultraviolet-B Radiation Flux in Alpine Regions. *Science* 248:206–208, 1990.

22. Caldwell, M. M., Robberecht, R., and Flint, S. D., Internal Filters: Prospects for UV-Acclimation in Higher Plants. *Physiol. Plant* 58:445–450, 1983.

23. Flint, S. D., Jordan, P. W., and Caldwell, M. M. Plant Protective Response to Enhanced UV-B Radiation Under Field Conditions: Leaf Optical Properties and Photosynthesis. *Photochem. Photobiol.* 41:95–99, 1985.

24. Markham, K. R., Franke, A., Given, D. R., and Brownsey, P. Historical Antarctic Ozone Level Trends from Herbarium Specimen Flavinoids. *Bull. Liasion du Groupe Polyphenols* 15:230–235, 1990.

25. Manabe, S., and Wetherald, R. T. Reduction in Summer Soil Wetness Induced by an Increase in Atmospheric Carbon Dioxide. *Science* 232:626–628, 1986.

26. Kellogg, W. W., and Zhao, Z.-Ci. Sensitivity of Soil Moisture to Doubling of Carbon Dioxide in Climate Model Experiments. Part I: North America. *J. Climate* 1:348–366, 1988.

27. Adams, R. M., Rosenweig, C., Pearl, R. M., Ritchie, J. T., McCarl, B. A., Glyer, D. J., Curry, R. B., Jones, J. W., Boote, K. J., and Allen, L. H., Jr. Global Climate Change and U.S. Agriculture. *Nature* 345:219–224, 1990.

28. Blasing, T. J., Stahle, D. W., Duvick, D. N. Tree Ring-Based Reconstruction of Annual Precipitation in the South-Central United States from 1750 to 1980. *Water Resources Res.* 24:163–171, 1988.

29. National Academy of Sciences. *Saline Agriculture: Salt Tolerant Plants for Developing Countries.* National Academy Press, Washington, DC, May 1990, No. 41-89-9.

30. White, R. M. The Great Climate Debate. *Sci. Am.* 263:36–43, 1990.

31. Schlesinger, M. E., and Jiang, X. Revised Projection of Future Greenhouse Warming. *Nature* 350:219–222, 1991.

CHAPTER 5

APPROACHES TO ENERGY EFFICIENCY

Energy is the key. The energy contained in fossil fuel became the force driving the wheels of industry and permitting rapid transportation by land, sea, and air. The world became manageable, people were brought closer together, and modern society was born. But not without a price: the liberation over the past 200 years of immense quantaties of carbon dioxide. It is time to move away from CO_2-emitting fuels. Over the past decade, suitable alternatives have become available. Within the next 25 years, both nonfossil fuels and carbon-efficient fuels could substantially decrease emissions of CO_2 by as much as 35% without the requirement for major technological innovations or breakthroughs. The technology is at hand. And such levels of reduction would significantly slow the rate of warming.

But change is difficult. The dismal record of the past five presidential administrations, Nixon, Ford, Carter, Reagan, and Bush, almost a quarter of a century, bears witness to that. A comprehensive national energy policy or strategy may never emerge. If energy independence continues to wax and wane with the price of oil, precious lead time to energy efficiency will most assuredly be lost. But waiting for the government to act may be unnecessary.

While change will require exceptional motivation and a range of incentives, a responsible public—individuals, businesses, corporations, the academic world—a public educated to the need for change can do those things that should be done and need to be done before complete knowledge of the climate system becomes available. But unfortunately, from the point of view of urgency, global warming has an inherent flaw. The time involved—"sometime in the next century," "50 years from now," or "midway into the 21st century"—mitigates against action now. Given a public tied to yearly cycles—the school year, vacations, birthdays, and other anniversaries, and income tax—along with professional politicians whose primary concern is getting elected or reelected for a 2-, 4-, or 6-year term, concern for 50 years from now when most will be retired or dead has little urgency compared to issues requiring attention now: AIDS, drug abuse, crime in the streets, unemployment hazardous waste. Undeniably, these are important issues but global warming, should it occur—uncertainty is yet another reason for lack of urgency—could pose far greater problems for society.

The American public has a history of response to challenge. The war in the Persian Gulf has focused the public's mind at a level not seen since World War II. And with the war successfully culminated, the time for change may be propitious; while the oil wells burn and troops remain in the Middle East, and the public is focused on oil.

An adequate supply of oil is a requirement. But diversity is the key to efficiency. To be effective, energy efficiency must include a variety of forms. Thus the following discussion examines a broad spectrum of appropriate preventive measures, which together can slow the accumulation of greenhouse gases. This is a direct approach to delaying, even averting, further warming. Let us then look at the diverse sources of energy.

THE CAR

We Americans love our cars. Our love affair with motor vehicles in general is known the world over. Americans love the speed, the freedom, the interiors, and especially an automobile front end. It is the rear that receives the least attention, or more to the point, the exhaust gases leaving the tailpipe that are of little or no concern. Thus it may come as a surprise to learn that fully 25% of the total annual accumulation of greenhouse gases the world over are contributed by motor vehicles. Closer to home, we Americans consume 40% of all the gasoline produced. That is, we Americans, 4% of the total world population, consume 40% of the world's supply of gasoline while the remaining 96% of the world's people divide the remaining 60% among them. We are the world leaders in gasoline consumption and, with it, emissions of CO_2. Thus it seems reasonable that, especially in the United States, motor vehicles could lead the way to emission reduction.

Automobiles are a central and vital part of American life. They have done well for us. But as the single greatest contributor of greenhouse gases, it is also clear that the automobile threatens our way of life. Yet they need not be abandoned or replaced. The solution is not that drastic. In fact, it is relatively benign. We must wean our cars from gasoline and provide them with a new source of power. And this new source of power is not out there somewhere waiting to be discovered. It is available, it has been tested, and it is ready for use: we need only ask for it. We need only to make it known and clear to automobile manufacturers that we want to convert, and the marketplace—ever alert and responsive to the wishes of consumers—will provide. It is as simple as that. And by switching fuel, by dropping gasoline, in favor of methanol, we can become independent of foreign oil, reduce the threat of greenhouse warming precipitously, and substantially reduce such air pollutants as carbon monoxide, ozone, and particulates. Air quality would rise to unprecedented heights. All this can be achieved almost at a stroke simply by switching fuels. It that is not motivation enough, financial incentives are also possible to assist in converting older model cars. Here again, we need only make this need clear to our political representatives at the state and federal levels. But what is this fuel of which we can expect so much?

Methanol is the simplest member of a class of organic compounds known as alcohols. Some are drinkable. Methane is not. Nor is methanol a modern invention. It was discovered by an Englishman, Robert Boyle, as long ago as 1661. It is a clear, colorless, volatile liquid that burns with a colorless flame and forms explosive mixtures with air. This latter characteristic recommended it early on as suitable for fueling internal combustion engines. But for the longest time, its higher price relative to gasoline kept it from becoming the fuel of choice—except for drivers at the "Indy 500," who use it almost exclusively to power their sophisticated racing cars. In addition to its higher octane rating of 100, compared to 90–97 for gasoline, which allows engines to run at higher compression ratios, and thus more efficiently, methanol is cleaner burning, which assures the additional benefit of longer engine life.

Its current "pump" price of $1.10–1.20 per fuel equivalent of a gallon of gasoline (it takes 2 gallons of methanol to drive a car as far as 1 gallon of gasoline will take it) means that it can now compete favorably with gasoline and with that singular development automobile manufacturers have designed and begun production of the engines. In fact, the U.S. EPA has given General Motors approval to manufacture a methanol-powered Chevrolet Lumina, the model GM expects to sell in California in 1991. But GM is not alone. The Ford Motor Company is producing methanol-powered engines for each of its weight/size car classes, and Chrysler and most foreign manufacturers plan to have these new cars available by 1993. The race to corner the American market is driven by the stringent air pollution standards adopted by California in an attempt to finally bring it smog and haze under control. It is the type of incentive manufacturers needed to initiate the move away from gasoline, which cannot possibly achieve the low level of exhaust emissions demanded by the standards. Methanol-powered cars can meet the standards. Thus it is that tough air pollution standards may well be the vehicle whereby carbon dioxide, the most conspicuous greenhouse gas, is simultaneously brought under control. The chemical equations that follow show quite clearly the contribution of CO_2 made by methanol and gasoline to an overburdened atmosphere:

$$\text{methanol: } 2CH_3OH + 3O_2 \rightarrow 2CO_2 + 4H_2O;$$

$$\text{gasoline: } 2C_8H_{18} + 25O_2 \rightarrow 16CO_2 + 18H_2O.$$

As shown in Figure 1, the high octane and low heat loss of the methanol engine allow the use of a smaller, more efficient engine, gas tank, and other parts. The cooling system could also be made smaller by switching from a heavier water-cooled system to a lighter air-cooled one. These changes would decrease the size and aerodynamical drag of the front end, resulting in a smoother riding car. And drivers looking for peppier cars will not be disappointed. Both the compact and mid-size models have been shown in proven ground tests to have increased acceleration, moving from 0 to 60 miles per hour in 11 seconds or less.

Because of current uncertainty over the availability of adequate supplies of methanol, as well as the lack of an in-place distribution network, automobile manufacturers will be offering flexible-fuel vehicles or dual-fuel delivery systems, designed to use either gasoline or methanol or a mixture of both. The new cars will be equipped with electronic or optical sensors, which will assess the type of fuel or mixture in the fuel line and, with the air of an on-board computer, instantaneously adjust fuel injection and spark requirements.

Dual fuel engines include bifuel systems and variable fuel systems. The bifuel engines have separate tanks for either methanol or gasoline, while the variable fuel system can burn a mixture of gasoline and methanol or either. Recently, the Chrysler Corporation developed a Methanol Concentration Smart Sensor, making possible the production of a true flexible-fuel vehicle that will be indistinguishable from the traditional gasoline system in terms of performance, durability, and reliability.

The "smart sensor" permits the use of gasoline and up to 85% methanol without any noticeable difference. Many of the flexible-fuel vehicles currently being tested are in fact alternative or dual-fuel systems with the capability of running on either gasoline, methanol, or a specific combination of the two, not a continual variation of the two. At the smart sensor's core is a built-in microprocessor that includes all the electronic circuitry required to process the sensor's information and relay analog output signals to the engine's computer. As shown in Figure 2, the smart sensor fits inside a nickel-coated plastic housing approximately the size of an audio cassette tape—an unusually compact unit. It is mounted in the fuel supply line between the fuel tank and the engine. As the fuel enters the sensor, the engine computer is instantaneously (10–50 milliseconds) alerted to the specific ratio of gasoline to methanol. The computer calculates spark advance and fuel injector width necessary to provide the correct amount of fuel to the engine.

In comparison to optical sensors, a microprocessor eliminates the risk of fogging as well as compensating for water/methanol mixtures, which at below-zero temperatures form a milky consistency, which can confuse an optical sensor. With advances such as this, alternatively fueled vehicles are being brought closer to reality.

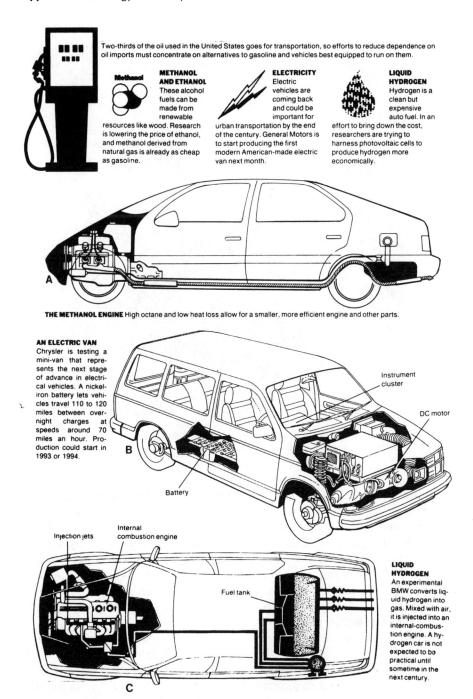

Two-thirds of the oil used in the United States goes for transportation, so efforts to reduce dependence on oil imports must concentrate on alternatives to gasoline and vehicles best equipped to run on them.

METHANOL AND ETHANOL These alcohol fuels can be made from renewable resources like wood. Research is lowering the price of ethanol, and methanol derived from natural gas is already as cheap as gasoline.

ELECTRICITY Electric vehicles are coming back and could be important for urban transportation by the end of the century. General Motors is to start producing the first modern American-made electric van next month.

LIQUID HYDROGEN Hydrogen is a clean but expensive auto fuel. In an effort to bring down the cost, researchers are trying to harness photovoltaic cells to produce hydrogen more economically.

THE METHANOL ENGINE High octane and low heat loss allow for a smaller, more efficient engine and other parts.

AN ELECTRIC VAN Chrysler is testing a mini-van that represents the next stage of advance in electrical vehicles. A nickel-iron battery lets vehicles travel 110 to 120 miles between overnight charges at speeds around 70 miles an hour. Production could start in 1993 or 1994.

Instrument cluster

DC motor

Battery

Internal combustion engine

Injection jets

Fuel tank

LIQUID HYDROGEN An experimental BMW converts liquid hydrogen into gas. Mixed with air, it is injected into an internal-combustion engine. A hydrogen car is not expected to be practical until sometime in the next century.

Figure 1. New fuels and new cars to use them: (A) a methanol engine, (B) an electric van, and (C) a hydrogen gas engine. (Reproduced with permission of *The New York Times*.)

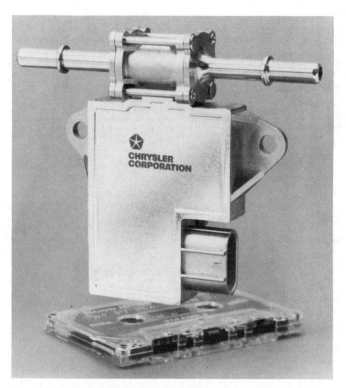

Figure 2. Chrysler Corporation Methanol Concentration Smart Sensor allows any fuel mixture of gasoline and up to 85% methanol with no noticeable difference in performance or driveability. The entire system, which includes a built-in microprocessor, fits inside a nickel-coated plastic housing about the size of an audio cassette.

Environmentally, these benefits, as important as they are for consumer acceptance, may not be as important as the fact that some critics see the use of great amounts of trees for methanol production as a loss of an important CO_2 link (although the major source of methanol will probably be the microbiologic digestion of wood pulp). However, plans being developed at the Oak Ridge National Laboratory foresee the widespread planting and harvesting of trees much like crops, but on a 6–8-year cycle, in which the young growing trees would absorb CO_2 offsetting any contribution made by methanol-powered vehicles.

As good as they are planned to be, methanol-powered cars will have competition. Prospective car buyers may prefer electric cars; a wise choice since the optimum way to reduce urban air pollution and with it CO_2 emissions is to curb it at its source. Electric cars will do this. As electrically powered cars eliminate all emissions, we can anticipate precipitous declines in nitrogen oxides, hydrocarbons, and CO_2. If the number of electric vehicles supersedes conventional internal combustion vehicles in large numbers, the hoped for delay in predicted global warming could become a reality. It is unfortunate that the state of the art is not 2–3 years ahead of itself.

Although electric cars were used on a limited basis in the 1930s for local deliveries, their lack of range and durability rapidly made them obsolete. Their revival has come as a consequence of major advances in battery development and current storage capacity. One of the great advantages of an electric car is that it uses no energy when it is idling in contrast to a conventional vehicle, which continues to do so. This takes on greater importance as traffic density increases. To their disadvantage, the range of electric cars is limited by the current state of battery technology. The minivan, whose battery configuration is shown in Figure 1B, represents the next major advance in battery technology. A nickel–iron battery generates sufficient power to permit travel at speeds up to 70 miles per hour for approximately 100–120 miles. Nickel–iron batteries are more durable than the conventional lead–acid type, which are used to start conventional engines, but they are also much more expensive to produce. Consequently, available electric cars can be double the price of conventional cars.

General Motors recently unveiled its Impact, a two-passenger car which it claims can accelerate from 0 to 60 miles per hour in 8 seconds and run 125 miles without a charge. Its lead–acid battery would surely require replacement annually, if not sooner, at considerable expense. The fact remains that until batteries with sufficient demonstrable power to maintain speeds of 55–75 miles per hour for 120–175 miles without recharging, and which are small enough and light enough to be packaged into a compact or mid-size model, become available, sales of electric cars will languish. On the other hand, California's stiff air pollution standards, which other states are certain to emulate, will force the wider use of electric cars however cumbersome by an early date. Service industries with large fleets will pioneer their use quickly. But the day of wider use is not far off. It is widely anticipated that by 1999, battery-powered cars will begin to make an impact.

In February 1991, General Motors (GM), Ford, and Chrysler entered into an agreement setting up a consortium, which the Department of Energy (DOE) and the Electric Power Research Institute (EPRI) will join. The U.S. Advanced Battery Consortium, as they call themselves, has as its goal the development of sophisticated battery technology to power electric vehicles, (EVs), but also to develop a hybrid electric vehicle (HEV). With the passage of California's new air pollution standards, which automobile manufacturers must view as onerous, the handwriting was indeed on the wall. California has yearly sales of some 2 million new cars and light trucks—serious numbers. The new standards virtually mandate that 2% or 40,000 vehicles be either electric powered or otherwise contribute zero pollutants to the air by 1998. This "quota" would rise to 10% or 200,000 vehicles by the year 2003. For automobile manufacturers the message was clear: produce a nonpolluting vehicle that the public will accept or lose the market to the Japanese or Europeans. The issue was nothing less than survival.

With some 185 million registered internal combustion engine vehicles already on the road, and 6–7 million new ones joining them yearly, the appeal of alternatively fueled or propelled cars and trucks is extremely compelling. In the more than 100 urban areas with chronically poor air quality, reducing the number of gasoline-driven vehicles by even a small percentage would dramatically improve air quality, because Evs and HEVs are so overwhelmingly less polluting. According to recent DOE figures, 63% of the 17 million barrels of oil consumed daily across the country are used for transportation, and of that 59% or 6 mbd are used to fuel passenger cars, vans, and light trucks. Consequently, each electric-type vehicle coming onto the road reduces the demand for oil by as much as 20 barrels per day. The arithmetic is simple and direct; the solution is clear.

The EPRI, producer of the G-Van, a battery-powered, fully electric standard-size van designed for fleet service, has shifted its emphasis to a hybrid electric vehicle (HEV)*, which appears to have substantial potential as an interim alternative until such time as fully functional batteries become available. The reasoning behind the HEV is that a combination of an electric drive system, storage batteries, and a small internal combustion engine should achieve better results than either a standard gasoline engine or any of the currently available electric drive systems.

The hybrid has the unique characteristic essential for control of emissions, that is, reducing the duration and peak magnitudes of the high emission levels occurring with cold starting. Cold starts are the source of 90% of the undesirable emissions. If drivers allowed their cars to idle for 30–60 seconds after starting, a major source of emissions would disappear. Apparently even 30 seconds is too long a wait for today's impatient motorists.

Hybrids appear to be the means of obtaining ranges close to that of gasoline engines. They combine the elements of more than one type of propulsion system; that is, the vehicle receives a portion of its energy as externally generated electricity. From a functional point of view, HEVs can be classified as series hybrids, parallel hybrids, and series/parallel hybrids. Figure 3 illustrates these pathways. The series hybrid delivers all the engine's power as well as the battery's power to the drive wheels by way of the same electric propulsion drivetrain. Thus the engine output is first converted to electricity. The parallel hybrid delivers engine power and battery output to the wheels via separate paths or by way of the same mechanical drivetrain, while the combination series/parallel hybrid permits reconnection in either of the two modes. Localized emissions (pollu-

---

* I am indebted to Dr. Alvin Salkind, College of Engineering, Rutgers University, for bringing me up to date on hybrid electric vehicles.

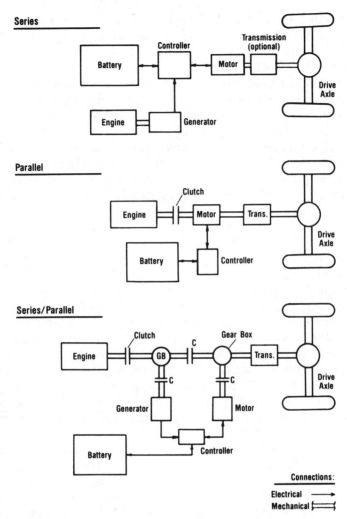

Figure 3. Simplified schematic illustrations classifying hybrid electric drivetrains by three major pathways. The series hybrid delivers all power to the drive wheels via the same electric drivetrain. Engine output is first converted to electricity. The parallel hybrid delivers power to the wheels via separate pathways or the same mechanical drivetrain, while the series/parallel hybrid permits combinations of either. (Courtesy of the Electric Power Research Institute, Palo Alto, CA.)

tion) and noise are all but eliminated if the HEV is operated in an all-electric mode.

EPRI has moved to develop a prototype hybrid version of the G-Van that will have twice the G-Van's range. A 6.5-kW, 10-hp on-board generator will provide supplemental direct-current motor power when required to slow the rate of battery discharge, thereby attaining longer ranges and extended time between charges. If initial tests go well, it is expected that commercial production of an extended range vehicle could begin by 1993. Given that HEVs do not qualify as zero pollution vehicles, regulators of California's air quality will be drafting new regulations to accommodate these low-emission transition vehicles.

This is a development that bears watching. If development is successful, and there is every expectation for it to be, and if drivers and companies can be motivated to purchase them in place of gasoline-driven vehicles, auto-mobile manufacturers will be appropriately motivated, and the country will be well on its way toward a substantial reduction of CO_2 emissions. Furthermore, the infrastructure of service and maintenance for hybrids is well into the planning stage. Five years ago few would have bet on this possibility. Today EVs and HEVs are a good bet to alter the conditions of greenhouse warming.

HYDROGEN

Hydrogen is an ideal fuel. When burned, its only emission product is water. Although, as shown in Figure lC, liquid hydrogen is being tested as another alternative to gasoline, it has only one-sixth the energy of gasoline; consequently, at this stage, a vehicle fitted with a liquid hydrogen delivery system would have an extremely limited cruising range. Additionally, its method of production, either from the electrolysis of water or via solar cells, makes it an exceptionally expensive fuel. Further experimentation is necessary to reduce its cost and increase its range before it can become competitive. Nevertheless, hydrogen cannot be ruled out. Fuel cells may well hold an answer.

A fuel cell converts hydrogen gas directly into electricity, heat, and water without combustion. It can be thought of as a continuously operating battery with an external fuel tank. The principle was first advanced in the late 1830s, but it remained dormant until the space program revived the idea, providing power for both the Gemini and Apollo missions. Currently, three fuel cells provide the on-board electricity for the space shuttles. As shown in Figure 4, the fuel cell is a curiously simple device, operating at near room temperatures and having no moving parts; thus it works silently. Essentially, it consists of a plastic membrane with carbon electrodes on both sides sandwiched between graphite plates. The active component of a solid polymer fuel cell consists of two porous catalyzed electrodes sepa-rated by an ion-conducting polymer membrane. This membrane is a thin, flexible, plastic sheet, resembling the clear plastic used to wrap foods. The electrodes are bonded to either side of the membrane electrolyte to form a membrane–electrolyte assembly. This is a clear advance over the caustic liquid electrolyte, such as potassium hydroxide, used in the Gemini and Apollo spacecrafts.

Figure 4. The Ballard fuel cell: a 1.7-kW, 20-cell hydrogen/air power section. (Courtesy of the Ballard Company, Vancouver, British Columbia.)

The fuel hydrogen, or a gas mixture containing hydrogen, passes through one of the porous electrodes, the anode. The reaction between the anode and the electrolyte creates an electric current. Hydrogen ions, which are electrically charged atoms, then flow toward the cathode—the positively charged pole—where oxygen (or air) reacts with hydrogen to form water and heat, which together produce steam. The electric current can be supplied as either direct or alternating current. And the fuel cell is environmentally benign with hydrogen being obtained from such clean sources as wind or solar energy.

With no moving parts in a fuel cell generator, a motor vehicle containing them would be nearly silent. And with small stacks stored throughout the car, there would be a major reduction in the size of the engine compartment under the hood. Vehicles could then be designed for people rather than engines. Accordingly, Ballard Systems together with the city of Vancouver and British Columbia Transit are working to develop hydrogen fuel cells for use in city buses. Expectations are that the first prototype will be available by December 1991. Their goal is to have a fleet of 15 passenger buses ready for 1994 when the Commonwealth Games take place in Vancouver. These buses will convey the athletes and other participants from site to site. Following this test, buses will be placed in regular city service to evaluate them against conventional buses. In addition, General Motors and the U.S. Department of Energy and Ballard are working to develop a hydrogen-powered passenger car. Again, it is the strict standards imposed by California's air pollution regulations that have become the catalyst, driving the wider use of imaginative innovations such as fuel cells. For example, a fuel cell the size of a standard household air conditioner unit generates sufficient energy to heat a home and supply its electric needs. The uses are unlimited: residential, industrial, automotive, and the generation of power at remote locations. Fuel cells will play a major role in reducing CO_2 emissions.

NATURAL GAS

Natural gas—methane—may be the key to energy independence for the United States, while simultaneously abating greenhouse gas emissions. Methane and methanol are different, but they are also similar. Methanol, with but a single carbon atom, is the simplest of organic alcohols, and methane, with a single carbon atom in its molecule, is the simplest of hydrocarbons. Methane can be a source of methanol and both will be contributing to the reduction of greenhouse gases.

Methane is a naturally occurring gas, which, like coal, was formed from decaying vegetation that was tightly packed under great pressure for tens of millions of years. In fact, it is the immense quantities of methane in coal deposits that will be used to reduce the use of coal and oil. The Federal Bureau of Mines estimates that hundreds of billions of cubic feet of methane (CH_4) escapes from coal beds annually to add to the atmospheric burden of yet another IR absorbing gas. Now, with new drilling techniques collecting it, the methane will be made available for both home heating and as a motor vehicle fuel. It is this latter use that we want to pursue here.

Natural gas has long been used for cooking and home heating. But in British Columbia, one of Canada's most forward-looking provinces, thousands of cars are fueled by it. Thousands of passenger cars, trucks, and taxis have been powered by methane for the past 7 years in a test to determine if natural gas is economical, safe, and convenient. After the Middle East oil embargo and the crises of 1979, the Canadian government decided to embark on a program that would free them of dependence on foreign oil—well before the atmospheric greenhouse was shown to be increasing in its content of constituent gases. The government, understanding that powerful incentives would be needed to change entrenched behavior and to get motorists to give methane a chance, and with natural reserves of some 1200 trillion cubic feet of methane on hand (an 80-year supply), baited the hook. The price of methane was set at 25 cents. That would buy an amount equivalent to 1 gallon of gasoline. For 25 cents a gallon, one could load up with as much methane as one's "gas" tanks could carry. The conventional internal combustion engine can readily run on methane without modification. They bit. People will alter their behavior—if the price is right. But the government didn't stop there. They persuaded the gas companies to provide low-cost financing for the tanks, which can be rented on a monthly basis. They also provided grants to defray the cost of converting to dual-fuel delivery systems. And to cap it off, traditional gas stations were induced to provide pumps similar to standard gas pumps. Currently, 50 stations throughout the province offer natural gas. Although many motorists have converted to the flexible-fuel system, few now use gasoline. Natural gas has additional benefits. With an octane rating of 130 compared to 90–97 for gasoline, methane is far better for high compression engines, which further increases efficiency, giving increased mileage and performance. And this cleaner fuel prolongs the life of the engine, spark plugs, and lubricating oil. As a result, engine life can easily be extended to 200,000–400,000 miles. Unheard of with gasoline. Furthermore, in the cold Canadian winters, there is no trouble starting because the fuel is already in a gaseous state.

Freedom from foreign oil can of itself be a worthy and satisfying accomplishment for the country. But methane offers an accompanying environmental benefit. By converting the billions of cubic feet of coal bed methane to CO_2, far less of it is emitted to the atmosphere, where it is more IR absorbing than CO_2. And the CO_2 that leaves the tailpipe of methane-powered automobiles is much lower than that contributed by gasoline. Consequently, there is a net loss of potentially warming gases to the atmosphere.

The British Columbia experiment has worked. It has proved itself and should be expanded. Natural gas is not a temporary measure; it is a fuel whose time has come.

MASS TRANSPORTATION

In the movie L.A. Story, the opening scene has Steve Martin driving to work but stuck in a larger-than-life traffic jam, getting nowhere. He's got to be at the radio station where he is a weatherman in 10 minutes, but the backup is sure to last for hours. He's in the far right lane. What does he do? He pulls out of the lane off the highway and onto the sidewalk, drives down the sidewalk, across lawns, through backyards, down long flights of steps to streets below, and through alleyways, arriving at work with 2 minutes to spare. No one seemed to mind. It was a regular occurrence. Perhaps the idea was a bit overdrawn, but it did highlight a common and notorious problem to which the audience could easily relate. We've all tried to avoid the jam-ups and bottlenecks by leaving earlier, staying later, and/or driving the back streets; often to no avail. We are the victims of too many vehicles packed into too little space. Too many two-wage earner families, each driving to work in separate cars; too many people living too far from their places of work; too many businesses and corporate headquarters moving from the cities to the suburbs; and too many people driving to work alone have all conspired to overwhelm our roads and highways. The problem is especially acute in many metropolitan areas on both the East and West Coasts, the regions that have reaped the greatest economic growth over the past 15 years. And the problem is additionally severe in cities such as Los Angeles, which lack a mass transportation network. There must be a better way.

Regional transportation authorities may be an answer. Patterned on the model of Port and Sewage Authorities, they have the mandate to bypass the meddlesome political interventions so common at local, county, and state levels, which so often impede decisionmaking because of conflicting interests of constituents. Are roads to be widened or added? Are there already too many? Perhaps a rapid transit system would be more appropriate. Should it be a high-speed magnetically levitated system that does not contribute to air pollution and greenhouse gas emissions? It would be up to the Authority to decide how best to move large numbers of people quickly, safely, and economically. These issues often stall on the question of who pays the bill, which in a recessionary period usually means that frustrated drivers will continue to be stuck in traffic for some time to come—primarily because priorities are inverted. The fact is that states and communities know best what they need locally. The federal government cannot dictate which type of system is best for any area even though federal money (tax dollars) covers 90% of the total cost of a project.

Relief may be in sight. Recently, the U.S. Senate by an overwhelming majority, 91 to 7, approved an overhaul of the federal transportation system. The new legislation gives the states primary decisionmaking authority for determining how the $123 billion allocated for transportation between 1992 and 1996 would be spent. No longer would states be in thrall to the DOT. Over $20 billion of the money package could be used for mass transportation. An opportunity is at hand. But cities will need to pull their acts together. Political wrangling is a luxury whose time is past.

Half the population of the United States lives in cities with 1 million or more people. If these cities and their increasingly snarled suburban traffic cannot function adequately, we all suffer.

Cars are multiplying faster than people. In 1990, the population of the United States was 250 million, and motor vehicle registration stood at 183.6 million, which means there were 1.36 people for each registered vehicle. Ten years earlier there were 1.63 people for every car. If this rate of increase continues, by the year 2020 there should be almost one car for every member of our population. At the point, we'll have national gridlock and the L.A. Story will become a reality. But it is not the traffic jams, the lack of movement, in and of themselves that are the sole cause for concern. It is rather the tremendous quantities of gasoline and oil burned that is so environmentally distressing. The answers are not to widen roads or to cut more roadways through the countryside; the answer lies in mass transportation. That's the priority. The U.S. Department of Transportation studies show that one bus can eliminate 35 automobiles, one street car can dispense with 50 passenger cars, and one train removes 1000 cars from the highway. Mass transportation can reduce the traffic jams, the air pollution, and the unconscionable levels of CO_2 emitted along the packed highways in the scramble to get from home to work and back. Recessionary periods notwithstanding, the public and elected officials must reorder their priorities. Mass transportation must move to the top of everyone's list. And there is yet another unconscionable by-product that mass transportation can alleviate—death on the highways.

Between 1975 and 1990, close to 800,000 people were killed in motor vehicle collisions—I was about to say accidents, but for the better part of 99% they cannot be described as accidental. Purposeful is closer to the facts. In addition to the deaths, there were millions of injuries: almost 6 million in 1990 alone. The damage to property and the economic losses over the 15 years runs into hundreds of billions of dollars. It is a staggering panoply of losses about which few seem concerned. However, from the evening that General Washington and his ragtag band of tattered troops crossed the Delaware River to attack the British at Trenton and Princeton, New Jersey, to General Schwartzkoff's masterly maneuvers in the War in the Persian Gulf, 662,572 members of the armed forces have lost their lives in all our wars. Apparently, it is safer to be a frontline soldier than a participant in the battle on the highways. It is also a fact of life—or death—that Americans believe they are not only safe but excellent drivers, needing no further regulation. But with some 50,000 men, women, and children killed on our highways each year, the appropriate description must be slaughter. And this slaughter is considered acceptable by our citizen-drivers. Accordingly, no political constituency can be expected to arise to urge a halt to this carnage on the roads because to do so would be to opt for self-regulation. Other than cancer, heart disease, and stroke, no other condition or illness exacts so heavy a toll. Motor vehicle deaths have become a highly effective method of population control. Mass transportation, a low priority item for the current and past administrations, could reduce energy expenditure and oil and gas use and save countless lives at one stroke. That ought to be motivation enough.

NUCLEAR POWER

On a global basis, nuclear power generates nearly 20% of the world's electricity and emits no carbon dioxide in the process. Obviously then, nuclear generated energy can make a significant contribution. Considering that worldwide electric power generation is the greatest single source of carbon dioxide emissions, a shift to nuclear power can substantially lower levels of CO_2. That is an acknowledged fact. Unfortunately, public concern over the safety of nuclear power generating plants, more than likely undue concern, jeopardizes its future acceptance.

Nuclear power has been moribund in the United States for more than a decade. The last order for a generating plant was placed in 1978, the year before the accident at Three Mile Island terrified the public. Nevertheless, the need for a nuclear power option is growing as the Persian Gulf crisis and yearly increases of atmospheric CO_2 underscore.

To split atoms and release the tremendous amounts of energy they contain, specially designed vessels or reactors are required. In most of these conventional containment vessels, water is circulated through the fuel core. The water extracts heat from the core, keeping it from overheating, while transferring the heat outside, where it produces steam that drives the turbines which ultimately generate electricity. Since water boils at 212°F (100°C), it must be kept under pressure in a vessel/reactor, or it will boil away and expose the core. This means that the reactor must be contained in a larger vessel. This vessel within a vessel concept is the basis for all existing plants generating power from atomic nuclei. In a conventional steam-generating plant, coal or oil is used to boil water, which is then converted to steam. A turbine, connected by a drive shaft to a generator, is driven at high speed as the steam strikes its blades and rotates them rapidly, enabling the generators to produce electrical energy. A nuclear power plant can be thought of in much the same way, except for the method of heating water to steam.

The heat in a nuclear plant is produced by the process of fission, which can occur in certain types of nuclear fuel. Essentially, a nuclear power generating plant consists of six components: reactor, core, fuel elements, coolant, moderator, and control rods.

A reactor is the part of a plant where the chain reaction occurs and where heat is generated. At the center of every nuclear reactor is the core, composed of nuclear fuel. When a neutron strikes an atom of fuel, it can be absorbed. This produces instability in the now heavier atom, which proceeds to split into two lighter atoms, as fission products. Heat (energy) and two or three neutrons are also produced. At this point, the reaction becomes chainlike in that huge numbers of neutrons begin striking other atoms, producing a shower of fissions.

An optimum chain reaction can be sustained in uranium if the neutrons are slowed prior to striking the fissionable fuel. This is accomplished by surrounding the fuel by a moderator, which absorbs a portion of the energy of the neutrons as they are released during the process. Several different materials are suitable as moderators, including water and graphite. Each of these moderators is the basis for a specific class of reactor.

The heat from the fission process is removed from the core by a continuous stream of fluid called the coolant. The heat in the coolant can be used directly to produce electricity.

Uranium-235 is the isotope most commonly used. A neutron striking an atom of ^{235}U is absorbed. The ^{235}U becomes unstable and splits, yielding two lighter fission products, heat and neutrons.

Only a few elements fission easily enough to be used in a nuclear power plant. Uranium is probably the most suitable. But it must be processed before it can be used. After refining, it is shaped into small cylinders or fuel pellets. Each are less than half an inch (12 mm) in diameter but each can produce the energy equivalent to 120 gallons of oil.

Fuel pellets are stacked in hollow tubes some 12 feet long. When full, these "pins" or "rods" are grouped together in bundles, the so-called fuel assembly. Within the assembly, rods must be carefully spaced to allow a liquid coolant to circulate freely between them. Approximately 200 nuclear assemblies are grouped together to make up the core of a reactor. It is this nuclear fuel in the core that generates the heat, just as the combustion of oil heats water in a boiler.

Interspersed among the fuel assemblies are movable control rods made of neutron-absorbing materials. It is upon the control rods that the speed of the chain reaction depends. Heat production is therefore moderated or augmented by inserting or withdrawing the control rods.

The entire reactor core, containing fuel assemblies and control rods, is enclosed in a stainless steel reactor vessel. And to further ensure safety, the steel-enclosed vessel is housed in a reinforced concrete structure.

Liquid coolant is pumped into the reactor through the core to remove heat. It is then pumped out of the reactor and used to produce steam. Those nuclear power plants using water as the coolant are known as light water reactors (LWRs). In this case, light water means ordinary water and it serves a dual function of moderating neutrons and transferring the heat of the reaction. Two distinct types of LWR are currently in use: boiling water reactors (BWRs) and pressurized water reactors (PWRs). In the United States, there are 84 nuclear reactors with operating licenses: 66% are PWRs, 27% are BWRs, and 7% are HWRs (heavy water reactors). In the boiling water reactor, pressure is controlled to ensure that the water boils as it passes through the core. consequently, steam is generated directly from the heat of the core. These are "direct cycle" systems with no intermediate steps. The steam produced drives a conventional turbine, which runs a generator.

That describes nuclear power generation through the 1980s, when new inherently safe designs emerged in response to growing public disenchantment and discomfort with the current generation of reactors. The PIUS (process inherent ultimately safe) reactors, not yet available, are designed to ensure safety even if the reactor is subjected to human error, equipment failure, or natural disaster.

In these reactors, personified by the Westinghouse 600 magawatt advanced passive light water plant, a prestressed concrete reactor vessel with a range of unique internal fittings replaces the conventional pressure vessel, the emergency core cooling system, containment shell, and spent fuel storage ponds. It effectively eliminates the possibility of a reactor core meltdown and simplifies plant design. Nor does it require the use of any mechanical or electrial components in any emergency calling for shutdown and after heat cooling for a period of more than 1 week for a range of possible conditions. The design criteria are to be met "without significant sacrifices in the economic and reliable performance as a generating plant" (1). As shown in Figure 5, the AP600 design places makeup tanks of borated cooling water (boron is a strong neutron absorber and automatically interrupts a chain reaction) above the reactor.

As illustrated, gravity alone is sufficient to convey adequate makeup water to the core to manage small leaks. If a large leak, such as a loss of coolant, occurred, gravity flow would be abetted by gas-pressure-driven

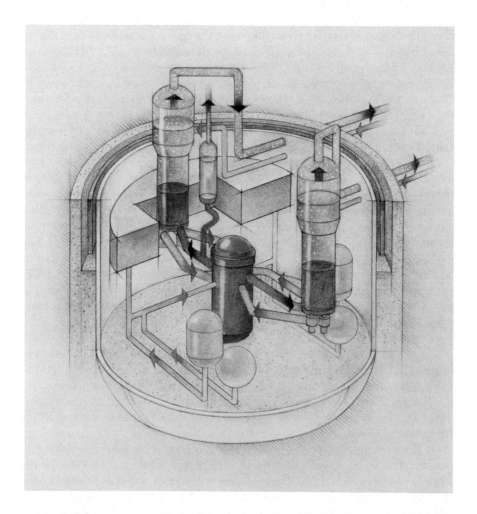

Figure 5. Basic design of an advanced passive 600-megawatt nuclear power generating plant. The essential feature is simplicity. (Courtesy of Westinghouse Electric Corporation, Pittsburgh, PA.)

fluid injection. The air-operated valves, which send this makeup water to the core, open automatically if they lose either pressure or their control signal. In the event of a loss of cooling water, additional water flows to the reactor from the large refueling water storage tank positioned above the reactor. Consequently, pumps are not required, and many of the components traditionally associated with safety and emergency cooling are eliminated. Simplicity is the rule (2). Thus the AP600 has no need of safety-grade coolers, chillers, pumps, or diesels. And the operation is no longer dependent on a technician to assure the prompt availability of makeup water in the event of either small or large leaks.

For the nuclear steam supply system, the new PIUS has a compact and simplified main loop configuration that enhances its operability, safety, and durability. These include a low power density core, canned reactor coolant pumps mounted on steam generators in a close-coupled configuration, and long-radius-bend reactor coolant pipes, which reduce pressure drops as well as the number of welds in the system. As shown in Figure 5, this loop arrangement has been simplified to the point that there are only two cold legs and one hot leg per loop. Clearly, 30 years of experience with first-generation reactors will provide a far safer second generation. With safety and simplicity come economic benefits; these new reactors will cost much less than their predecessors.

In addition to the PIUS designs, an experimental Intergral Fast Reactor (IFR) is nearing completion under the supervision of the Argonne National Laboratories. It requires the use of liquid sodium in place of water as the cooling liquid. Liquid sodium boils at 1600°F (857°C), which permits its use at normal pressures. Since the vessel containing the reactor can therefore be much smaller, the entire system is submerged in a pool of liquid sodium. Thus all the potential cooling fluid, not just that flowing through the core, is available in the event of an accident. The liquid sodium carries the heat to water outside, where the steam to drive the turbines is produced—as in conventional systems.

Another change is the fuel itself. Where conventional reactors use uranium oxide, the IFR uses an alloy composed of 75% uranium, 15% plutonium, and 10% zirconium. And it uses a new fuel-processing technique that relies on electrorefining rather than chemical extraction. Since electrorefining requires smaller and simpler machinery, it will reduce the amount of radioactive waste. With a simpler processing system in the plant along with the reactor, the risk of accident or theft during transportation is also reduced.

If the IFR proves equal to the claims being made for it, nuclear power may become the remedy for potential global warming. For example, each 1000-megawatt electric power reactor prevents the emission of 6 million tons of CO_2 per year, the level produced by coal-fired plants producing an equal amount of electricity. Together, the world's 430 nuclear power plants prevent some 1.6 billion tons of CO_2 emissions, fully 8% of total emissions. If that figure could be doubled, the greenhouse warming would probably no longer be an issue. Consider this from another vantage point. A single 1000-megawatt (MW) coal-fired electrical generating plant requires over 2 million tons of coal per year, and that amount of coal must be hauled by the equivalent of 20,000 coal cars via a system that is itself heavily energy intensive. An oil-fired plant uses 10 million barrels of oil a year, which requires more than six supertankers for its transportation. Mark you, that's one plant. By comparison, a typical uranium fission plant requires only the equivalent of one rail car load of uranium dioxide per year, and when the first fusion plants come on-line, the amount of fuel they require, 1000 pounds, can be carried easily in a pick-up truck. But the environmental benefits do not end there. The same coal-fired plant also produces enough ash (solid waste) to fill 2000 railroad cars, which must be hauled to ever-diminishing land-fill sites for burial. In addition, about 140 tons of sulfur dioxide and nitrogen oxides per day are vented to the atmosphere. Oil has no ash but emits similar amounts of sulfur and nitrogen compounds. Fission and fusion plants have no comparable emissions. These comparisons should be given dispassionate consideration if we are serious about reducing greenhouse gases. However, until the nuclear power industry shows that its plants can be operated safely, nuclear power will not emerge from its ashes, as it were. Until everyone takes nuclear power seriously, which means understanding its potential benefits as well as its potential harm, the battle against global warming will be waged without one of its most potent weapons.

For those whose opinion of nuclear generated electric power as an appropriate source of energy may have altered as a consequence of the accident at Chernobyl, in the Soviet Ukraine, perhaps a measure of comfort can be taken from the recent report by the UN's International Atomic Energy Agency (IAEA), which found that although hundreds of thousands of Soviet citizens suffer stress and anxiety because of the accident, in fact, there is no evidence of widespread leukemia or thyroid illness. "There were," they stated, "many important psychological problems of anxiety and stress related to the accident." But, they continued, "these were wholly disproportionate to the biological significance of the radioactive contamination." Thus the IAEA study found most Soviet children to be generally healthy and that the health of adults was no different in areas affected or unaffected by the explosion. Although, unfortunately, there was and will be illness and death as a consequence of the accident, for the most part, like Three Mile Island, it was a media event blown out of proportion to the actual effects. As Hans A. Bethe, a Nobel Laureate in physics, recently commented, an accident of the Chernobyl type "can not occur in the United States." Americans should not be afraid of nuclear power.

SOLAR POWER

The surface of the sun can be considered a nuclear power plant or furnace. Here energy is created by nuclear fusion, which releases tremendous amounts of energy as heat and light. As detailed in Chapter 2, solar radiation is emitted from the sun's surface with an energy distribution similar to that of a black body, or perfect emitter, at a temperature of approximately 6000 K (°C). The value of the solar constant, the radiation intensity received per second by 1 square meter of the upper atmosphere, is 1.353 W/m^2 (1.95 langleys/minute or 428 Btu/foot^2/hour). If this value is multiplied by the circumference of the earth and then by 24 (hours) and then by 365 (days), the yearly output is found to be 1.49 × 10^{18} kwh/yr; that is, 85,000 times more than the yearly amount of electricity used by the entire world's population. Obviously, even allowing for highly inefficient collection systems, no more than 20–30%, the amount of available potential energy is both tremendous and clean. That's the key. In less than 1 hour, enough solar energy reaches the earth to meet the world's needs for a year. The challenge is not simply to harness solar energy, although that of course is fundamental, but to make it available in areas distant from points of collection. That is, from sunny to less sunny areas, both day and night, clouds or no clouds.

A widely used method for taking advantage of the available energy allows the sun's heat to strike fluid-filled tubes contained in roof-top collector panels. As the fluid circulates throughout a building, a home or office, it radiates warmth for space heating. It is also used to heat water. In another process, sunlight falls upon solar cells—photovoltaic cells—and the incident energy is then converted directly into electricity without any mechanical movement or polluting by-products, and these fuel cells, as described earlier, are more efficient than conventional power plants at converting fuel to power. Solar energy is clean, renewable, and a highly promising alternative to fossil fuel.

However, in order for solar energy to take its place alongside other serious alternatives to fossil fuel, it would be necessary to find ways to store it and move it from its site of production to areas where it is needed. Converting solar energy to electricity is only a partial solution to this problem, because storing electricity is expensive and inefficient. A better way would be to convert the collected energy to a chemical form that can be stored and transported to places where it is needed. Until recently, that has been only a good idea. Now, however, Israeli scientists appear to have solved the transportation and storage problems. If their experiments prove successful, the world could be weaned from fossil fuel. The idea is stunning.

Led by Moshe Levy of the Weizmann Institute's Energy Research Center, the Israelis have concentrated on gathering the sun's energy to make steam rather than electricity. The concept that has been moved along is that of the "chemical heat pipe," which is based on the reversible reaction by which methane is reformed into hydrogen and carbon monoxide:

$$CH_4 + CO_2 \rightarrow 2H_2 + 2CO + \text{heat } (250 \text{ kJ/mole}).$$

The reforming reaction is conducted in a solar furnace at 800–1000°C, in which methane and steam are reacted. The gaseous reaction products are cooled, stored at high pressure, and transported to the point of use. The stored solar energy is released in a "methanator" plant, which brings hydrogen and carbon monoxide together, producing methane and heat used to generate steam, electric power, or both—as shown in the equation. The methane is returned to the solar collection site, completing the cycle, as shown in Figure 6. Thus far, the solar chemical heat pipe has been operated as a closed loop by Levy's group and their experiments have proved that the cycle can be repeated indefinitely with no emissions of CO_2 into the atmosphere. With no greenhouse contributions, and the fact that the gas can be stored for use when and where needed, this work could be the key needed to lead the world away from the use of fossil fuel.

Additional research at the Weizmann Institute has demonstrated that solar heat can be used to drive other high-temperature reactions. One of these has shown that oil shale can be gassified by solar heat, obtaining a high yield of fuel. These reactions can be used to produce hydrogen, which may become the energy staple of the 21st century.

WIND POWER

Wind-derived energy is not new. Sails have been used for thousands of years and windmills for about 500. Hence wind power is possibly the earliest example of energy conversion. However, it wasn't until the 1980s

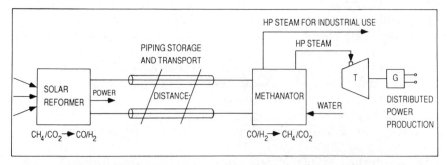

Figure 6. System for conversion of solar to chemical energy, and its storage and conversion. (Courtesy of Moshe Levy, Energy Research Center, Weizmann Institute of Science, Rehovot, Israel.)

that wind power became economically competitive for generating electricity. Since 1980, the cost of 1 kilowatt/hour of wind-generated electricity dropped to less than 7 cents. Compared to the new generation of coal-fired plants, which can produce power at 5–6 cents per kilowatt/hour, wind power is suddenly high competitive.

Energy derived from wind is coming of age as a nonpolluting, renewable source. Over the past decade, more than 16,000 wind turbines have come on-line in California and Hawaii. During this same period about 1700 megawatts of wind-generated electric capacity were established around the world. Recall that 1 megawatt is equal to 1 million watts. A generator with its 28-foot rotor blades mounted atop an 80-foot open truss tower is shown in Figure 7. Some 7500 of these are in operation at Altamont Pass (Figure 8), the site of one of the world's largest "wind farms." From a single workstation, an operator can monitor several thousand turbines simultaneously while scanning detailed operating data on individual turbines. In California alone, wind power generates approximately 1500 megawatts of power, which translates to over 1% of the state's total commercial energy

Figure 7. Workers atop an 80-foot open truss tower installing the rotor that will drive the turbine's generator. (Courtesy of U.S. Windpower, Livermore, CA.)

Figure 8. Panoramic view of the "wind farm" at Altamont, California, with its thousands of operating turbines. (Courtesy of the Electric Power Research Institute, Palo Alto, CA.)

budget: a quite respectable contribution by a relatively new and novel source, which further reduces the need for fossil fuel.

The basic principles of wind turbines, which combine 28-foot long rotor blades, derived in part from helicopter technology, with small induction generators, are relatively uncomplicated. The essential components of the turbine, within the nacelle, generator, control system, transmission, and rotor, are shown in Figure 9. However, the technology is maturing rapidly and newer variable speed rotors are being introduced, as are 54-foot-long blades that will sweep an area over three times that of the smaller blade; these advanced units will also have three times the power rating—300 kW. Variable speed turbines allow the rotor and generator to speed up with gusting or stronger winds. The increased rotational energy is converted into additional electricity without increasing torque on the drivetrain. The convertor, however, maintains a constant-frequency line output despite the generator's variable output frequency. Predictions are that, by the end of the 1990s, some 3000 megawatts of wind-generated electric power will be contributed to the state's power supply. And with the number of high-wind passes in California, the price per kilowatt will be further reduced and highly competitive. Interestingly enough, in both the San Francisco

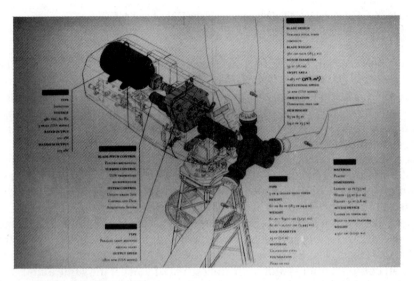

Figure 9. Essential components of a turbine shown in cutaway view. (Courtesy of U.S. Windpower, Livermore, CA.)

and Suison Bay (Solano County) areas, the utility demand peak is higher in summer when the wind is stronger, and at Suison the daily wind power pattern dovetails nicely with the utility's late-afternoon peak-demand period. Of course, the greater the amount of time a turbine is running and generating, the more energy it produces and the lower is the cost of each kilowatt/hour generated. Because winds do not blow steadily anywhere, levels of performance or capacity factor—the proportion of time the unit is generating—will affect costs. In the early 1980s, capacity factors at many wind farms ran below 15%. Now, stations such as Altamont manage 35%, which is near the maximum for California's wind resources.

The major factors involved in gain or loss of capacity factors is siting error and height of rotor blades. Siting errors of as little as 10 feet can reduce energy production by as much as 10%. Similarly, assessment of wind speed, direction, and height has further contributed to more efficient power generation. Obviously then, America's wind power resources will have much to say about the potential expansion of wind power nationally. For example, across the country average wind speeds ranges from 12 to 21 miles per hour. It is the exceptional area that averages 16–21 mph, but a number, for example, Hawaii, California, Oregon, Washington, Idaho, Montana, North Dakota, South Dakota, New Mexico, and Arizona in the West and Southwest, along with Maine, Vermont, and New Hampshire in the Northeast, average 12–14 mph. Thus there is substantial room for expansion; but again, incentives will be necessary to give this nonpolluting, renewable resource the impetus needed to increase its use.

ENERGY CONSERVATION: AN INDIVIDUAL AND NATIONAL COMMITMENT

Ordinarily, we don't think of it in these terms, but as of the 1990 population census, the United States had 250 million people; a quarter of a billion. Of course, China with over 1 billion has more, but a quarter of a billion is a lot of people—that's the key. Although there are many things each of us can do to use energy more efficiently, in terms of global warming, individual efforts may seem inconsequential. Not so. With so many of us our efforts take on staggering proportions. Consider this. How many light bulbs do you have in your house? Think about it. I've asked droves of people the same question and none knew. Ask someone that question, preferably in their home. The number is always larger than the quesstimate. Take a room by room count. You've probably never done this before and you are in for a surprise. Say you have 35; 35 incandescent light bulbs of various wattages. By simply replacing one 75-watt bulb with the new 18-watt compact fluorescent bulb, not only will you get the same amount of light provided by the 75-watt, but the new bulb will last 10–13 times longer and cut your lighting bill more than half. And that's not all. Over the longer life of the bulb there will be a savings in coal-fired electric energy equal to a half-ton of carbon dioxide. That's the point. Now imagine the reduction in CO_2 emissions if every household in the United States replaced only one bulb. But what if all the bulbs in all the houses and offices were replaced. Now let your imagination really soar. Imagine the all but incomprehensible energy savings on an international level if all of eastern Europe—Poland, Romania, Hungary, Czechoslovakia, Bulgaria, the Soviet Union, and what recently was East Germany—India, and China all converted from standard incandescent light bulbs to the more energy efficient compact fluroescents. All these countries have low efficiency bulbs, poorly functioning and low efficiency coal-fired power plants, and large grid losses in their electrical distribution networks, and all use low-grade coal. Can you imagine the beneficial consequences to the atmosphere? The point is simple but trenchant. People have tremendous power for change. Let us now think in terms of single individuals acting alone. We are in this together. And together the problem can be surmounted reasonably and with little discomfort.

Energy efficiency means that people do more with the same—or possibly less—amount of energy without lowering living standards. That's the essential point. Improved energy efficiency will also protect the environment, reduce our dependence on foreign oil, and promote economic growth. And there are many easy and inexpensive ways to reduce the use of energy without sacrificing comfort. To begin with, you have only to ask yourself how you keep cool, stay warm, how much and what kind of light you need for reading, word processing, playing Ping-Pong, lighting your home or office, how you cook, wash and dry clothes, and how you pack your refrigerator. There are ways to save energy in just about everything we do. Lighting is, of course, the major consumer of electricity in the United States, accounting for 25% of all energy use. But incandescent bulbs (IBs) do more for heating the air than lighting it. Only 10% of the generated electricity passing through an IB is converted to visible light. They are immensely inefficient and thus wasteful of the fuel producing their energy. Compact fluorescents convert better than 30% of their electrical energy to light. And they use only one-fourth the wattage to emit the same soft light as incandescents. How can that be? Fluroescent lighting, not all that new on the lighting scene, has garnered a bad reputation for the poor quality of light it provided. People look pale and interiors washed out. Fortunately, the new compacts have remedied this and are now richer in the red, green, and blue wavelengths, providing a soft light that compares favorably to incandescent. But it is necessary to read labels carefully. Bulbs are rated from 0 to 100 on a Color Rendering Index (CRI). The higher the value, the truer the color rendition. The best compacts have a CRI over 80. But there are problems.

Compacts are larger than standard incandescent bulbs, come in a wide variety of wattages and styles, and are far more inexpensive. Because demand for them remains light, they are not widely available and range in price from $20 to $40 each. Herein hangs a tale.

Earlier, I raised the question of the number of bulbs in a house. Obviously, that can vary from 5 to 100. But with prices such as these, changeover can be expected to be slow. Few families would be inclined to invest $400–500 upfront on the expectation of future reductions in their electric bills, nor is greenhouse warming seen as sufficiently critical. For less affluent households, such outlays could not be budgeted. Incentive programs are needed. In fact, many public utility companies are responding to the challenge. Some are giving away bulbs; some are offering their customer steep reductions to get them to switch. Why not. It is in their best interest to do so. They will continue to sell energy to light the compact fluorescents, but they will save large sums in construction costs. The costs of building power plants and ancillary equipment such as substations and powerlines are enormous: costs that are ultimately passed on to the consumer. Incentives are needed. If we are going to be weaned away from energy inefficiency to efficiency, easing the pain of the initial costs will surely reduce the changeover time.

Energy efficiency cannot be limited to homes. Large savings can be achieved by businesses retiring old cooling systems, motors, and lighting equipment and replacing them with more efficient proven technologies or installing energy efficient equipment as new buildings are constructed. To motivate these changes, utility companies are offering financial incentives in the form of sizable cash rebates once qualified equipment has been purchased and installed. Rebates are also available for the installation of steam air conditioning equipment as an alternative to electric cooling, as is air conditioning equipment fueled by natural gas. And rebates will be provided to reduce the purchase and installation costs of the equipment. Businesses and residential buildings with appropriate storage space can chill or freeze water at night and store it to air condition their buildings the following day. This involves circulating air over the cooled water or ice and distributing the cool air throughout the building. Cool storage helps avoid peak summer daytime demand for power. Along the rebates for installing the necessary equipment, savings accrue from avoiding peak power periods. Energy efficient motors can also reduce energy usage and operating costs in a broad range of commercial and industrial applications. Again, rebates are available for installing the qualifying equipment.

The need for an increasingly large supply of electric power is nowhere more evident than on the thousands of college and university campuses across the country. The approximately 4 million resident post-high-school students with their bewildering collections of power-demanding appliances have forced the rewiring of dormitories and residential apartments to accomodate the increased demands for energy. It is the odd student who doesn't claim ownership of a personal computer and printer, TV, VCR, popcorn poppers, refrigerator, oven, microwave, hair-dryer, ham-radio, CD player, electric shaver, stereo system, coffeemaker, alarm clock, and answering machine. Many have heat lamps, toasters, and radios. To these can be added the need for close support of air conditioning, photo copiers, and fax machines. Many students do not have the remotest idea where the energy comes from or how it is generated—other than placing a plug n a wall outlet. Yet they would be among the first to march or sit for global warming. That they are mighty consumers of fossil fuel is lost on them. And they would be among the most difficult to wean from their "toys." The fact that the universities, colleges, and private secondary schools accommodate the load of appliances has much to say about their priorites. Many schools have had to boost their energy capacity by 50% over the past 5 years and the future suggests demand will continue to grow. Apparently the need for power is growing by 3–5% per year on many campuses, at a time when student enrollment has reached a plateau. By comparison, increases in energy needs in cities such as New York and Los Angeles have risen 1.7 and 1%, respectively: two to three times less than the student enclaves. Conservation is not yet "in" among students, nor do university and college administrations think in these terms. Their response to the array of electronic equipment coming into their residence halls is to add additional lines and increase capacity. The use of natural gas, compact fluorescent bulbs, high-efficiency lamps and ballasts, and energy efficient motors is yet to appear.

But energy conservation for its own sake can be misrepresented. Senator Timothy E. Wirth, sponsor of energy efficiency legislation, and one of the most vigorous advocates of strong measures to forestall what he sees as the calamity of global warming, seriously believes that, "what we've got to do in energy conservation is try to ride the global warming issue even if the theory of global warming is wrong." He informs us further that "to have approached global warming as if it is real means energy conservation, so we'll be doing the right thing anyway in terms of economic policy and environmental policy" (4). Perhaps. Unfortunately there is also the problem of falsely portraying, and thereby further confounding the comprehension of the warming issue by sending the wrong messages to an already confused and frightened public sorely in need of the best available information.

While energy conservation is unquestionably a worthwhile endeavor, to fraudulently represent global warming in order to promote greater energy conservation is malicious mischief. Above all, integrity on the part of our elected officials is what's currently needed. As a nation, we've paid a heavy price over the past twenty years for the loss of trust in government. Double speak and double think, on the part of our leaders, in the name of seemingly great virtue cannot be condoned.

The alternatives to oil are limited only by the imagination. In southern California's Imperial Valley, a power plant fueled by cow manure is generating enough electricity to supply as many as 20,000 homes and, in the process, saves 300,000 barrels of oil that otherwise would be needed. It is the first commercial plant that burns only cattle chips for fuel. The 15-megawatt plant is located next to a cattle feedlot, where conveyor belts feed 40 tons of manure an hour into the plant. There it is dried and burned in special ovens. The heat produces more than 150,000 pounds of steam hourly to drive a turbine and electric generator. In addition to saving oil, the manure is no longer carted off and dumped in landfills. This type of steam generating plant could easily be multiplied dozens of times. With over 114 million head of cattle, 56 million hogs and pigs, and 11 million sheep advantageously concentrated around the country, there is more than enough manure to dispense with over 6 million barrels of oil a year. Such plants would also substantially decrease the pressure on landfills and could be good business. There's profit in manure.

Fighting global warming is good business, and opportunities for reducing energy use are widespread. Cheap oil has dulled the traditional American genius for inventiveness. The slumbering giant is about to awaken and flex some muscles. The results could be exhilarating. Fighting global warming may prove to be the nation's most important undertaking in half a century.

One of the most inspired and beneficial "muscle flexing" responses may not be all that far off. Down the road, perhaps 5—possibly 10—years, the entire process of photosynthesis (as shown in Figure 11, Chapter 4) will have been fully worked out in laboratories around the country and world. Scientists will be able to control and run the photosynthetic process without the need for green plants. With this astonishing capability, it will not only be feasible but practical to produce unlimited quantities of carbohydrates, hydrogen, and methane. With hydrogen emitting only water upon combustion, and methane emitting far less CO_2 than either coal or oil (gasoline), sources of fuel to power all types of engines (motors) would be readily available, freeing the country of dependence on foreign oil and obviating the need for coal. Once the entire process is on hand, it would then be only a matter of time for engineers to design and construct an operating facility. On that day, humankind will have arrived at the point of converting the sun's energy directly into liquid fuels by using the excess carbon dioxide in the atmosphere. If global warming can be held off or maintained at low levels until then, the entire problem may be solved for us. This type of research needs our fullest support.

The essence of this delineation of the numerous and germane approaches to reducing greenhouse gas emissions is to show that we are poised on a threshold, a point of departure whereon an armory of new technologies are about to become available nationally and internationally, along with those that will come on-line over the next decade, which together could substantially affect CO_2 emissions. That such is clear. And, if the most recent projections are correct, time to avert or delay global warming may well be on our side. If nothing else, it should also be evident that calamity and catastrophe should have no place in our thinking about and response to global warming. In the lexicographic sense of causing extreme havoc or misery, there is no calamity or catastrophe. If, by calamity and catastrophe, change is implied, than perhaps that may be, here and there. In general, however, people will go about their business, trying to make a go of life and living.

What will be difficult will be "selling" conservation on a broad scale. Perhaps the war in the Persian Gulf should have continued for another 6 months in order to focus the minds of the people a bit more permanently. As it is, the 100-hour ground war—the real war—may have made thoughts of oil only a temporary aberration. With so short a war, the price of oil tumbled rapidly and, with it, the need to replace oil, not because of price, but because of its contribution to a potentially warmer world. This is the point that is being missed or overlooked. Since the 1970s, with the first oil embargo, the roller coaster price of oil has played hob with even the most specious pretense at attempts to develop an alternative fuel program and independence from foreign oil. Oil has always been considered the fuel of choice. Any potential alternative is always compared to oil at its current price. If it had a higher price the alternative was rejected out of hand; if it had a lower price, there is a chance for acceptance. If the price of oil dropped, any chance was soon forgotten. And along with it, any thought of an alternative energy policy. Thus it is time to consider alternatives to oil for other reasons—the price of oil notwithstanding. But the price of oil is, in fact, artificial. The world is awash in oil. Even without Iraqi and Kuwaiti oil, there is so much oil available that the price per barrel could easily fall to $10 or less and would continue to earn billions of dollars for the oil states for whom the production of a barrel of oil costs less than $1. This, in itself, suggests that any comparison with alternative fuels is nonsensical and misleading. We need to forget the price of oil and concentrate on the CO_2 emission problem. If we are convinced that fossil fuel is indeed the culprit, then the price of oil is irrelevant. Until people in high places have the requisite information, that CO_2 emissions are affecting climate, oil will continue to be untouchable—well, perhaps not to everyone. As noted earlier, the public utilities have seen the light and understand that profits will continue to accrue, even as energy conservation is actively promoted. The U.S. automobile manufacturers could go either way. Theoretically, it is of no concern to them what type of fuel is used. They will sell cars if they are competitive. But—and here the chicken and egg metaphor is most applicable—they will not take the initiative to build methane or methane/ ethanol fueled cars if these fuels are not widely available, and the fuel producers will not produce the fuel without guarantees that sufficient vehicles requiring their fuel will also be available. And so it goes and will continue to go until a third party steps in to interdict this vicious cycle. And who might that third party be? Well, we already have more than intimations. Obviously, the problem could be solved in a moment by a decision at the federal level to shift from gasoline to another fuel and to do so in a reasonable way—say, 20,000—30,000 new cars per year, giving the fuel producers and distribution system ample opportunity to establish their networks. But that does not appear likely at this juncture.

A more likely "third party" could be the states, which possess the sovereignty, as we have seen in California, to establish motor vehicle exhaust emission regulations, the likes of which can only be met by fuels other than oil based. But one state cannot be a conversion force. It will take nine, the highly populous nine—California, New York, Illinois, Texas, Florida, Pennsylvania, Ohio, Michigan, and New Jersey. They are strategically located from border to border and sea to shining sea. The move would be irresistible. The automotive industry and fuel producers could not avoid being drawn in—not unless the automobile producers would like to post "going out of business" signs, because the Japanese, German, and Italian builders would fill the void instantly. Foreign automobile manufacturers abhor a vacuum. Here then is the place for a people's crusade. Each of those eight states—California is already there—needs to develop a network, a political network, which can carry the message to their state legislatures, a message that is unambiguous, a message that their elected representatives understand is the voice of the people. Thus nine pivotal states hold the key to realistic energy conservation: strict air pollution regulations. And fear of job loss need not be a cause for concern. The establishment of a new fuel distribution and dispensing network will bring with it an employment bonanza. The "grass roots" can literally change the face of America or, more to the point, can significantly alter the magnitude and direction of projected global warming. All it takes is the enactment of eight strict air pollution standards. If people are serious, it can't lose. It's simple, direct, and eminently workable. America doesn't really need Alaska's oil. Energy security does not mean more oil, even domestically produced. Energy security and attention to the possibility of global warming mean alternative fuels.

LET THE VIKINGS BE OUR GUIDE

Through the 99 years of the fifteenth century, as colder weather descended upon Europe and Greenland, the Viking's inflexibility to change assured their extinction. They could not adapt to their Inuit neighbors' type of clothing, style of boat, or style of life. They could not forsake cows for sheep or take up fishing. They were, first and foremost, Europeans. Four hundred years of living among the Eskimo had taught them nothing, and no matter how the environment changed, they would have no part of it. And so "finis" was written to their lives, a monument to inflexibility. That has never been an American failing. However, a new element may sorely try this national tradition.

We live in a period of "bottom lines" and short-term gains—quick profits to keep shareholders happy. Long-term planning, especially preventive planning, has never been our strong suit. In addition, while life expectancy has been steadily increasing since the turn of the century, the average for the entire population is currently 75; it is still relatively short. Thus concern for a potentially serious problem of uncertain consequences unfolding by the year 2030 or 2060, 40–60 years hence—half a lifetime for some, more than half for others—is totally foreign; it is not of the moment and thus lacks credibility and urgency. And without a strong political constituency, few if any politicians will spend their political coin or place their careers on the line for voteless issues. Let the politicians of 2030 and 2050 deal with their special problems. Unfortunately, that is muddled, self-serving, and short-sighted political thinking. But it is a fact of life. If this issue isn't joined now, we may squander our period of grace. On the other hand, "the public"—the American people—suggest unity, which may not exist. Until February 1991, an objective observer could have looked upon us and have seen a fragmented nation of individual groups concerned with promoting special causes or vested interests, talking at cross purposes. The war in the Gulf may have changed that, a consummation devoutly to be wished.

At this point, science and scientists can take us no further. They have given us all the data, and those data are good, as far as they go. But the data are replete with uncertainty; they must be, given our limited knowledge of the universe. Beyond is for the public. The public must arm its collective self in order to make the decisions that will surely affect it for hundreds, if not thousands, of years. Jean Piaget, the Swiss psychiatrist, understood the problem well. "The great danger today," he remarked, "is of slogans, collective opinions, ready made trends of thought. We have to be able to resist individually, to criticize, to distinguish what is proven and what is not." Indeed, people must arm themselves with appropriate information and they must convey their beliefs to their elected representatives, or they will fall victim to well organized special interest groups who have their own agendas and have little concern whether it squares with the national need.

The facts in the case notwithstanding, the mean annual surface temperature of our planet varies considerably from year to year and is an inescapable fact of life, unrelated to protestations of global warming. Nevertheless, as I believe I have shown, while there is reason to question the occurrence of any significant warming, its consequences could be so disagreeable for contemporary society that such a possibility must be given serious consideration. Unfortunately, however, for the general discussion, there has been a universal lack of candor. The "facts" in the case have been presented as a prediction, a *fait accompli*, when in fact they should be seen as Wigley, Jones, and Kelly have suggested, a "scenario", a guide to plausible patterns of climate change that could accompany a warming (3). Candor, in short supply, is currently what is needed.

Default is no way to solve pressing problems. But do people really want to share in decisionmaking? Haven't they sent their representatives to state houses and Washington to do that for them? Given the sophistication and complexity of the issues surrounding global warming, are they able to participate or is the problem intractable? I think not. Time and again we have seen that, when the public is given information in a form that it can digest, the right decision will be made—the one best for it. That's the key. Edmund Burke, the Dublin-born British statesman, was right. "The public interest," he said, "requires doing today those things that men of intelligence and good will wish, five or ten years hence, had been done." Why wait until it's too late?

REFERENCES

1. Hannerz, K. Towards Intrinsically Safe Light Water Reactors. ORAU/IEA-83-2(M)-REV. Oak Ridge Associated Universities. July 1983.

2. McCandles, R. J., and Redding, J. R. Simplicity: The Key to Improved Safety, Performance and Economics. *Nuclear Eng. Int.* 34(424):20–24, 1989.

3. Wigley, T. M. L., Jones, P. D., and Kelly, P. M. Scenario for a Warm, High-CO_2 World. *Nature* 283:17–21, 1980.

4. Stanfield, R. L. Less Burning, No Tears. National Journal. August 13th, 1988. pp. 2095–

GLOSSARY

aerosol   Particulate material, other than water or ice, in the atmosphere ranging in size from approximately 10^{-3} to larger than 10^2 μm in radius. Aerosols are important in the atmosphere as nuclei for the condensation of water droplets and ice crystals, as participants in various chemical cycles, and as absorbers and scatterers of solar radiation, thereby influencing the radiation budget of the earth–atmosphere system, which in turn influences the climate on the surface of the Earth.

albedo   The fraction of the total solar radiation incident on a body that is reflected by it, or the percentage of the total incident solar radiation that is reflected back into space without absorption.

anthropogenic   Produced or influenced by human activity.

aphelion   The point on the Earth's orbit at which the Earth is at its farthest distance from the Sun.

altithermal   A period of high temperature particularly the one from 8000 to 4000 b+p (before the present era), which was apparently warmer in summers, as compared with the present, and with precipitation zones shifted poleward.

atmosphere   The envelope of air surrounding the Earth and bound to it by the Earth's gravitational attraction. Studies of the chemical properties, dynamic motions, and physical processes of this system constitute the field of meteorology.

atmospheric window   The spectral region between 8.5 and 11.0 microns (μm) where the atmosphere is essentially transparent to long-wave radiation.

biogeochemical cycle   The chemical interactions that exist between the atmosphere, hydrosphere, lithosphere, and biosphere.

biomass   The total dry organic matter or stored energy content of living organisms that is present at a specific time in a

	defined unit (community, ecosystem, crop, etc.) of the earth's surface.
biosphere	The part (reservoir) of the global carbon cycle that includes living organisms (plants and animals) and life-derived organic matter (litter, detritus). The terrestrial biosphere includes the living biota (plants and animals) and the litter and soil organic matter on land; the marine biosphere includes the biota and detritus in the oceans.
biota	The animal and plant (fauna and flora) life of a given area.
black body radiation	The electromagnetic radiation emitted by an ideal black body; a perfect emitter.
C^3 plants	Plants (e.g., soybean, wheat, cotton) whose carbon fixation products have three carbon atoms per molecule. Compared with C^4 plants, C^3 plants show an increase in photosynthesis with a doubling of CO_2 concentration and less decrease in stomatal conductance, which results in an increase in leaf-level water-use efficiency.
C^4 plants	Plants (e.g., maize, sorghum) whose carbon fixation products have four carbon atoms per molecule. Compared with C^3 plants, C^4 plants show little photosynthetic response to increased CO_2 concentration above 340 ppm but show a decrease in stomatal conductance, which results in an increase in photosynthetic water-use efficiency.
CAM plants	Crassulacean aid metabolism. Plants (e.g., cactus and other succulents) that unlike the C^3 and C^4 plants have temporarily separated the processes of carbon dioxide uptake and fixation when grown under arid conditions. They take up gaseous carbon dioxide at night when the stomata are open and water loss is minimal. During the day when the stomata are closed, the stored CO_2 is released and chemically processed. When CAM plants are not under water stress they then follow C^3 photosynthesis.
carbon budget	The balance of the exchanges (incomes and losses) of carbon between the carbon reservoirs or between one specific loop (e.g. atmosphere–biosphere) of the carbon cycle. An examination of the carbon budget of a pool or reservoir can provide information about whether the pool or reservoir is functioning as a source or sink for CO_2.

carbon cycle   All parts (reservoirs) and fluxes of carbon; usually thought of as a series of the four main reservoirs of carbon interconnected by pathways of exchange. The four reservoirs, regions of the earth in which carbon behaves in a systematic manner, are the atmosphere, terrestrial biosphere (usually includes freshwater systems), oceans, and sediments (includes fossil fuels). Each of these global reservoirs may be subdivided into smaller pools, ranging in size from individual communities or ecosystems to the total of all living organisms (biota). Carbon exchanges between the reservoirs by various chemical, physical, geological, and biological processes.

carbon dioxide fertilization   Enhancement of growth or in the net primary production due to CO_2 enrichment that could occur in natural or agricultural systems as a result of an increase in the atmospheric concentration of CO_2.

carbon isotope ratio   Ratio of carbon-12 to either of the other, less common, carbon isotopes, carbon-13 or carbon-14.

carbon sink   A pool (reservoir) that absorbs or takes up released carbon from another part of the carbon cycle. For example, if the net exchange between the biosphere and the atmosphere is toward the atmosphere, the biosphere is the source and the atmosphere is the sink.

climate   The statistical collection and representation of the weather conditions for a specified area during a specified time interval, usually decades, together with a description of the state of the external system or boundary conditions. The properties that characterize the climate include thermal (surface air temperatures, water, land, ice), kinetic (wind and ocean currents, together with associated vertical motions and the motions of air masses, aqueous humidity, cloudiness and cloud water content, groundwater, lake lands, and water content of snow on land and sea ice), and static (pressure and density of the atmosphere and ocean, composition of the dry air, salinity of the oceans, and the geometric boundaries and physical constants of the system). These properties are interconnected by the various physical processes such as precipitation, evaporation, infrared radiation, convection, advection, and turbulence.

climate change   The long-term fluctuations in temperature, precipitation, wind, and all other aspects of the earth's climate.

External processes such as solar irradiance variations, variations of the Earth's orbital parameters (eccentricity, precession, and inclination), lithosphere motions, and volcanic activity are factors in climatic variation. Internal variations of the climate system also produce fluctuations of sufficient magnitude and variability to explain observed climate change through the feedback processes interrelating the components of the climate system.

climate sensitivity

The magnitude of a climatic response to a perturbing influence. In mathematical modeling of the present climate, a given parameter is changed and the difference between the resulting disturbed simulation and control simulation is a measure of the climate sensitivity to that disturbance.

climate signal

A statistically significant difference between the control and disturbed (see climate sensitivity) simulations of a climate model.

climate system

The five physical components (atmosphere, hydrosphere, cryosphere lithosphere, and biosphere) that are responsible for the climate and its variations.

climatic optimum

The period in history from about 5000 to 2500 B.C., during which surface air temperatures were warmer than at present in nearly all regions of the world. In the Arctic region, the temperature rose many degrees, and in temperate regions the increase was 1.0–1.7°C. In this period, there was a great recession of glaciers and ice sheets, and the melt-water raised sea level by about 3 meters.

cloud

A visible mass of condensed water vapor particles or ice suspended above the earth's surface. Clouds may be classified on their visual appearance or form of the cloud or cloud height.

cloud albedo

Reflectivity that varies from less than 10% to over 90% of the insolation and depends on drop sizes, liquid water content, water vapor content, thickness of the cloud, and the sun's zenith angle. The smaller the drops and the greater the liquid water content, the greater the cloud albedo, if all other factors are the same.

cloud feedback

The coupling between cloudiness and surface air temperature in which a change in surface temperature could lead to a change in clouds, which could then amplify or diminish the initial temperature perturbation. For example, an increase in surface air temperature could increase the evaporation; this in turn might increase the extent of cloud cover. Increased cloud cover would reduce the solar radiation reaching the earth's surface, thereby lowering the surface temperature. This is an example of negative feedback and does not include the effects of long-wave radiation and the advection in the oceans and the atmosphere, which must also be considered in the overall relationship of the climate system.

cryosphere   The portion of the climate system consisting of the world's ice masses and snow deposits, which includes the continental ice sheets, mountain glaciers, sea ice, surface snow cover, and lake and river ice. Changes in snow cover on the land surfaces are by and large seasonal and closely tied to the mechanics of atmospheric circulation. The glaciers and ice sheets are closely related to the global hydrologic cycle and to variations of sea level and change in volume and extent over periods ranging from hundreds to millions of years.

deforestation   The removal of forest stands by cutting and burning to provide land for agricultural purposes, residential or industrial building sites, roads, and so on, or by harvesting the trees for building materials or fuel. Oxidation or organic matter releases CO_2 to the atmosphere. Regional and global impacts may result from the release of CO_2 to the atmosphere at a rate similar to that for fossil fuel releases.

dendro-chronology   The dating of past events and variations in the environment and the climate by studying the annual growth rings of trees. The approximate age of a temperate forest tree can be determined by counting the annual growth rings in the lower park of the trunk. The width of these annual rings is indicative of the climatic conditions during the period of growth: wide annual rings signify favorable growing conditions, absence of diseases and pests, and favorable climatic conditions, while narrow rings indicate unfavorable growing conditions or climate. The most sensitive (variability in ring widths) tree ring chronologies come from trees whose growth has been limited in some way by climatic or environmental factors.

dendro-climatology   The use of tree growth rings as proxy climate indicators. Tree rings record responses to a wider range of climatic variables over a larger part of the Earth than any other type of annually dated proxy record.

energy balance models   Models in which explicit calculations of atmospheric motions are omitted. In the zero-dimensional models, only the incoming and outgoing radiation is considered. The outgoing infrared radiation is a linear function of global mean surface air temperature and the reflected solar radiation is dependent on the surface albedo. The albedo is a step function of the global mean surface air temperatures, and equilibrium temperatures are computed for a range of values of the solar constant. The one-dimensional models have surface air temperature as a function of latitude. At each latitude, a balance between incoming and outgoing radiation and horizontal transport of heat is computed.

flux   The emissions from the biosphere, including the emissions from sources and the take-up associated with emissions' sinks.

fossil fuel   Any hydrocarbon deposit that can be used as a fuel such as petroleum, coal, and natural gas.

GCM   General circulation model. Hydrodynamic models of the atmosphere on a grid or spectral resolution that determine the surface pressure and the vertical distributions of velocity, temperature, density, and water vapor as a function of time from the mass conservation and hydrostatic laws, the first law of thermodynamics, Newton's second law of motion, the equation of state, and the conservation law for water vapor. (Abbreviated as GCM.) Atmospheric general circulation models are abbreviated AGCM, while oceanic general circulation models are abbreviated OGCM.

GFDL   Geophysical Fluid Dynamics Laboratory.

glacier   A mass of land ice, which is formed by the further recrystallization of firn (compacted granular snow of the previous year). A glacier flows slowly (at present or in the past) from an accumulation area to an ablation area. Some well-known glaciers are the Zermatt, Stechelberg, Grindelwald, Trient, Les Diablerets, and Rhone in Switzerland; the Nigards, Gaupne, Fanarak, Lom, and Bover in Norway; the Wright, Taylor, and Wilson Piedmont glaciers in Antarctica; the Bossons Glacier in France; the Emmons and Nisqually glaciers on Mt. Ranier, Washington; Grinnell glacier in Glacier

National Park, Montana; the Dinwoody glacier in the Wind River Mountains and the Teton glacier in Teton National Park, both in Wyoming.

greenhouse effect
A popular term used to describe the roles of water vapor, carbon dioxide, and other trace gases in keeping the earth's surface warmer then it would be otherwise. These "radiatively active" gases are relatively transparent to incoming short-wave radiation but are relatively opaque to outgoing long-wave radiation. The latter radiation, which would otherwise escape to space, is trapped by these gases within the lower levels of the atmosphere. The subsequent reradiation of some of the energy back to the surface maintains surface temperatures higher than they would be if the gases were absent. There is concern that increasing concentrations of greenhouse gases, including carbon dioxide, methane, and chlorofluorocarbons, may enhance the greenhouse effect and cause global warming.

greenhouse gases
Those gases, such as water vapor, carbon dioxide, tropospheric ozone, nitrous oxide, and methane, that are transparent to solar radiation but opaque to long-wave radiation. Their action is similar to that of glass in a greenhouse, but the process causing warming in a greenhouse differs from that involved in the greenhouse effect. Thus the terms greenhouse effect and greenhouse gases are misnomers.

Holocene
The most recent epoch of the Quaternary period covering approximately the last 10,000 years.

hydrologic cycle
The movement and exchange of water among the earth, atmosphere, and oceans.

hydrosphere
The aqueous envelope of the earth including the oceans, freshwater lakes, rivers, saline lakes and inland seas, soil moisture and vadose water, groundwaters, and atmospheric vapor.

hypsithermal period
The period about 4000 to 8000 years ago when the earth was apparently several degrees warmer than it is now. More rainfall occurred in most of the subtropical desert regions and less in the central midwest United States and Scandinavia. It is also called the "altithermal period" and can serve as a past climate analog for predicting the regional pattern climate change should the mean earth surface temperature increase due to an in-

ice age   A glacial epoch or time of extensive glacial activity. Also written as Ice Age, which refers to the latest glacial epoch, the Pleistocene epoch.

ice and snow albedo   The reflectivity of ice- and snow-covered surfaces. The albedo of freshly fallen snow may be as much as 90%, while older snow may have values of 75% or less. The larger the areal extent of snow and ice cover, the higher the albedo value.

IGBP   International Geosphere Biosphere Program.

infrared radiation   Electromagnetic radiation lying in the wavelength interval from 0.7 to 1000 μm. Its lower limit is bounded by visible radiation, and its upper limit by microwave radiation. Most of the energy emitted by the earth and its atmosphere is not infrared wavelength. Infrared radiation is generated almost entirely by large-scale intra-molecular processes. The triatomic gases, such as water vapor, carbon dioxide, and ozone, absorb infrared radiation and play important roles in the propagation of infrared radiation in the atmosphere.

insolation   The solar radiation incident on a unit of horizontal surface at the top of the atmosphere. It is sometimes referred to as solar irradiance. The latitudinal variation of insolation supplies the energy for the general circulation of the atmosphere. Insolation depends on the angle of incidence of the solar beam and on the solar constant.

insolation   A contraction for incoming solar radiation; the solar radiation impinging on the earth.

isotope   One of two or more atoms that have the same atomic number (i.e., the same number of protons in their nuclei) but have different mass numbers.

lithosphere   The component of the earth's surface comprising the rock, soil, and sediments. It is a relatively passive component of the climate system and its physical characteristics are treated as fixed elements in the determination of climate.

Little Ice Age   A cold period that lasted from A.D. 1550–1600 to about A.D. 1850 in Europe, North America, and Asia. This period was marked by rapid expansion of mountain glaciers, especially in the Alps, Norway, Ireland, and Alaska. There were three maxima, beginning about 1650, about 1770, and 1850, each separated by slight warming intervals.

LLNL	Lawrence Livermore National Laboratory.
long-wave radiation	The radiation emitted in the spectral wavelength greater than 4 μm corresponding to the radiation emitted from the earth and atmosphere. It is sometimes referred to as terrestrial radiation or "infrared radiation."
model sensitivity	Determined by using alternate input values to the program and studying their effects on model outputs.
NOAA	National Oceanic and Atmospheric Administration.
pCO$_2$	The partial pressure of CO_2 in the atmosphere and the ocean. In the atmosphere, the partial pressure of CO_2 is defined as the pressure the CO_2 would exert if all other gases were removed. The sum of the partial pressure of all the atmospheric gases will equal the atmospheric pressure. The partial pressure of CO_2 in the atmosphere is determined by the atmospheric CO_2 concentration and atmospheric temperature. In the ocean the pCO$_2$ is determined by the amount of dissolved CO_2 and H_2CO_2. It varies with alkalinity, latitude, depth, and temperature. Biological processes in the ocean also exert an influence on the pCO$_2$ in the ocean.
perihelion	That point on the Earth's obit at which the Earth is closest to the Sun.
photosynthesis	The process by which green plants use light to synthesize organic compounds (primarily carbohydrates) from carbon dioxide and water, using the light absorbed by chlorophyll as the energy source. Oxygen and water vapor are released in the process. Photosynthesis is dependent on favorable temperature and moisture conditions as well as on the atmospheric carbon dioxide concentration. Increased levels of carbon dioxide can increase net photosynthesis in many plants.
Quaternary period	The last 2 million years of the earth's history. It is divided into two epochs: the Pleistocene—2 million years ago to approximately 100,000 years ago—and the Holocene—the period from approximately 10,000 years ago to the present. The Quaternary period is the artificial division of time separating prehuman and human periods. It contains five ice ages and four interglacial ages and temperature indicators seem to show sharp and abrupt changes by several degrees.
radiation balance	The difference between the absorbed solar radiation and the net infrared radiation. Experimental data show

that radiation from the earth's natural surfaces is rather close to the radiation from a black body at the corresponding temperature; the ratio of the observed values of radiation to black body radiation is generally less than one, but close to it.

radiative-convective models
Thermodynamic models that determine the equilibrium temperature distribution for an atmospheric column and the underlying surface, subject to prescribed solar radiation at the top of the atmosphere and prescribed atmospheric composition and surface albedo. Submodels for the transfer of solar and terrestrial radiation, the heat exchange between the eartearth's surface and atmosphere, the vertical redistribution of heat within the atmosphere, the atmospheric water vapor content, and clouds are included in these one-dimensional models.

radiative forcing
A phrase used to identify the relative contribution of a particular trace gas to the total climate change associated with all radiatively important trace gases.

radiosonde
A balloon-borne instrument for the simultaneous measurement and transmission of meteorological data up to a height of approximately 30,000 meters (100,000 feet). The height of each pressure level of the observation is computed from data received via radio signals.

respiration
A biochemical process by which living organisms take up oxygen from the environment and, in the case of plants, consume some of the photosynthate (organic matter produced by photosynthesis) that they have synthesized during daylight hours, or in the case of animals, release both carbon dioxide and heat during respiration.

solar constant
The rate at which solar energy is received just outside the Earth's atmosphere on a surface that is normal to the incident radiation and at the mean distance of the Earth from the Sun. The value is 0.140 watt/cm^2.

statistical–dynamical models
Models that treat the dynamical processes statistically by relating them parametrically to temperature and temperature gradients. The major differences between these models and the general circulation models (GCMs) are the degree and scale of the parameterized processes.

stoma (pl. stomata)
A minute pore in the epidermis of plant leaves or stems. Stomata, which are bordered by guard cells that regu-

late the size of the opening, function in gas exchange between the plant and the external environment. The stomatal apparatus or stoma consists of the stoma plus guard cells.

stratosphere   The region of the upper atmosphere extending upward from the tropopause (8–15-kilometer altitude) to about 50 kilometers. The thermal structure is determined by its radiation balance and is generally very stable with low humidity.

sunspot   A relatively dark, sharply defined region on the solar disk, marked by an umbra approximately 2000 K cooler than the effective photospheric temperature, surrounded by a less dark but also sharply bounded penumbra. The average sunspot diameter is about 3700 kilometers, but can range up to 245,000 kilometers. Most sunspots are found in groups of two or more, but they can occur singly. Sunspots are cyclic, with a period of approximately 11 years. The quantitative description of sunspot activity is called the Wolf sunspot number, denoted R. The Wolf sunspot number is also referred to as "Wolfer sunspot number," "Zurich relative sunspot number" or "relative sunspot number."

sunspot   A region on the sun's surface which is cooler than surrounding areas.

surface albedo   The fraction of solar radiation incident on the earth's surface that is reflected by it. Reflectivity varies with ground cover, and during the winter months it varies greatly with the amount of snow cover (depth and areal extent). Roughness of terrain, moisture content, solar angle, and angular and spectral distributions of ground level irradiations are other factors affecting surface albedo.

terrestrial radiation   The total infrared radiation emitted from the earth's surface. The atmosphere emits, absorbs, and transmits radiation, and the net flux of radiation at any one point depends on the distribution with height of temperature and water vapor. Terrestrial radiation provides a major part of the potential energy changes necessary to drive the atmospheric wind system and is responsible for maintaining the surface air temperature within limits for liveability.

troposphere   The inner layer of the atmosphere below about 15 kilometers, within which there is normally a steady decrease of temperature with increasing altitude. Nearly all clouds form and weather conditions manifest themselves within these region and its thermal structure is due primarily to the heating of the earth's surface by solar radiation, followed by heat transfer by turbulent mixing and convection.

USGS   United States Geological Survey.

water vapor   Water substance in vapor form and the source of all forms of condensation and precipitation. Water vapor, clouds, and carbon dioxide are the main atmospheric components in the exchange of terrestrial radiation in the troposphere, serving as a regulator of planetary temperatures via the greenhouse effect. Approximately 50% of the atmosphere's moisture lies within about 1.84 kilometers of the earth's surface and only a minute fraction of the total occurs above the tropopause.

weather   The day to day meteorological conditions in a local area.

WMO   World Meteorological Organization.

APPENDIX

Scale of Geological Time

Era	Period	Epoch	Evolutionary Events
Cenozoic	Quaternary	Recent (Holocene)	End of cold period and onset of warming trend: beginning and maturation of current civilization, tool makers, hungers, Little Ice Age, and industrial growth; carbon dioxide found to be rising.
		Pleistocene (2 mya to 100 kya)	Emergence of human beings in Africa, Europe, and Asia; major glaciations; grasslands expand
	Tertiary	Pliocene (7–2 mya)	Hominids appear; emergence of the Panamanian land bridge; North and South America united
		Miocene (26–7 mya)	Appearance of cobras
		Oligocene (38–6 mya)	Angiosperms dominant
		Eocene (54–38 mya)	Mammals, birds, and insects dominate the earth
		Paleocene (65–54 mya)	
	Cretaceous (145–65 mya)		K-T boundary; collision of India and Asia; dinosaur extinctions; dispersal of insects; angiosperms arise and expand, gymnosperms decline, flowering plants evolve, Rocky Mountains arise; North America connected with Europe; appearance of such "modern" trees as maple, oak, birch, and elm

Scale of Geological Time (*Continued*)

Era	Period	Epoch	Evolutionary Events
Mesozoic	Jurassic (200–145 mya)		Extinction of reptiles and amphibians; dominance of land by gymnosperms; birds evolve
	Triassic (245–200 mya)		Dinosaurs, turtles, crocodiles, and mammals appear
	Permian (285–245 mya)		Extinction of over 96% of all species; formation of supercontinent of Pangaea; great expansion of reptiles; decline of amphibians; last of the trilobites; appearance of cockroaches and scorpions
	Carboniferous (360–285 mya)		Great coal forests, dominated at first by lycopsids and sphenopsids, then ferns and gymnosperms; amphibians expand; first great dispersal of insects; reptiles evolve
	Devonian (430–360 mya)		Expansion of primitive vascular plants; seed plants appear (gymnosperms); fishes appear; first amphibians; insects abundant
Paleozoic	Silurian (500–430 mya)		Vascular plants arise toward end of period; scorpions emerge from sea to land; coral reefs emerge
	Ordovician (570–500 mya)		Marine algae abundant; first vertebrates; jawless fish; reef-building algae; clams and starfish appear
	Cambrian (700–570 mya)		First life appears; Primitive marine algae (e.g., Cyanophyta and Chlorophyta); stromatolite-building cyanobacteria; marine invertebrates abundant (including representatives of most phyla)
Precambrian	(3800–700 mya)		Primitive marine forms; algae and soft-bodied invertebrates; atmosphere begins to change and oceans form
Azoic	(4600–3800 mya)		Existence of planet Earth

kya, thousands of years ago.
mya, millions of years ago.

Abbreviations and Acronyms Commonly Used in CO_2 Research	
B.P.	Before Present
DOE	Department of Energy
GARP	Global Atmosphere Research Project
GCM	General Circulation Model
GFDL	Geophysical Fluid Dynamics Laboratory
GISP	Greenland Ice Sheet Program
GISS	Goddard Institute for Space Studies
GLAS	Goddard Laboratory of Atmospheric Sciences
GSFC	Goddard Space Flight Center
IGBP	International Geosphere–Biosphere Program
IGY	International Geophysical Year
IR	Infrared radiation
MLO	Mauna Loa Observatory, Hawaii
NAS	National Academy of Sciences
NASA	National Aeronautics and Space Administration
NCAR	National Center for Atmospheric Research
NCC	National Climatic Center
NOAA	National Oceanic and Atmospheric Administration
NODC	National Oceanographic Data Center
NRL	Naval Research Laboratory
NSF	National Science Foundation
ppmv	parts per million by volume (sometimes designated simply as ppm)
SIO	Scripps Institution of Oceanography
USDA	United States Department of Agriculture
USGS	Unites States Geological Survey
WDC	World Data Center
WMO	World Meteorological Organization

International System of Units (SI): Prefixes

Prefix	SI Symbol	Multiplication Factor
exa	E	10^{18}
peta	P	10^{15}
tera	T	10^{12}
giga	G	10^{9}
mega	M	10^{6}
kilo	k	10^{3}
hecto	h	10^{2}
deca	da	10
deci	d	10^{-1}
centi	c	10^{-2}
milli	m	10^{-3}
micro	μ	10^{-6}
nano	n	10^{-9}
pico	p	10^{-12}
femto	f	10^{-15}
atto	a	10^{-18}

Source: W. C. Clark ed. Carbon Dioxide Review: 1982, Oxford University Press, New York, 1982. (Reprinted with permission of Oxford University Press.)

Celsius (Centigrade)/Fahrenheit Temperature Conversion Chart

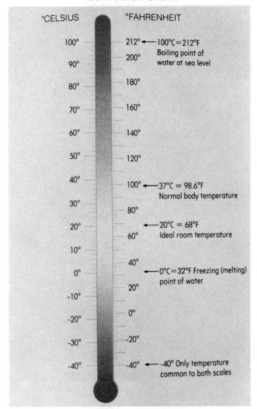

°CELSIUS | °FAHRENHEIT

100°C = 212°F
Boiling point of water at sea level

37°C = 98.6°F
Normal body temperature

20°C = 68°F
Ideal room temperature

0°C = 32°F Freezing (melting) point of water

-40° Only temperature common to both scales

Common Conversion Factors

Area–length–volume
  1 acre = 43,560 ft^2 = 4047 m^2
  1 acre-foot = 1.2335×10^3 m^3
  1 cubic foot (ft^3) = 0.02832 m^3
  1 hectare (ha) = 10,000 m^2 = 2.47 acres
  1 square mile (mi^2) = 2.59×10^6 m^2
Pressure
  1 atmosphere = 76.0 cm Hg = 1013 millibars (mb)
  1 bar = 0.9869 atmosphere
  1 pascal (Pa) = 0.9869×10^{-5} atmosphere = 1×10^{-2} mb = 10 μbar
    = 1.4505×10^{-4} pounds per squar inch (psi)
Factors for carbon and carbon dioxide
  1 mole C/liter = 12.011×10^{-3} Gt C/km^3
  1 ppm by volume of atmosphere CO_2 = 2.121 Gt C
  1 mole CO_2 = 44.011 g CO_2 = 12.011 g C
  1 g C = 0.083 mole CO_2 = 3.664 g CO_2

Adapted from W. C. Clark, ed. *Carbon Dioxide Review: 1982*, Oxford University Press, New York, 1982.
C = carbon.

INDEX